SWITCH/ROUTER
ARCHITECTURES

SWITCH/ROUTER ARCHITECTURES

Shared-Bus and Shared-Memory Based Systems

JAMES AWEYA

John B. Anderson, *Series Editor*

Published by John Wiley & Sons, Inc., Hoboken, New Jersey.
Published simultaneously in Canada.

No part of this publication may be reproduced, stored in a retrieval system, or transmitted in any form or by any means, electronic, mechanical, photocopying, recording, scanning, or otherwise, except as permitted under Section 107 or 108 of the 1976 United States Copyright Act, without either the prior written permission of the Publisher, or authorization through payment of the appropriate per-copy fee to the Copyright Clearance Center, Inc., 222 Rosewood Drive, Danvers, MA 01923, (978) 750-8400, fax (978) 750-4470, or on the web at www.copyright.com. Requests to the Publisher for permission should be addressed to the Permissions Department, John Wiley & Sons, Inc., 111 River Street, Hoboken, NJ 07030, (201) 748-6011, fax (201) 748-6008, or online at http://www.wiley.com/go/permission.

Limit of Liability/Disclaimer of Warranty: While the publisher and author have used their best efforts in preparing this book, they make no representations or warranties with respect to the accuracy or completeness of the contents of this book and specifically disclaim any implied warranties of merchantability or fitness for a particular purpose. No warranty may be created or extended by sales representatives or written sales materials. The advice and strategies contained herein may not be suitable for your situation. You should consult with a professional where appropriate. Neither the publisher nor author shall be liable for any loss of profit or any other commercial damages, including but not limited to special, incidental, consequential, or other damages.

For general information on our other products and services or for technical support, please contact our Customer Care Department within the United States at (800) 762-2974, outside the United States at (317) 572-3993 or fax (317) 572-4002.

Wiley also publishes its books in a variety of electronic formats. Some content that appears in print may not be available in electronic formats. For more information about Wiley products, visit our web site at www.wiley.com.

Library of Congress Cataloging-in-Publication Data is available.

ISBN: 978-1-119-48615-2

Printed in the United States of America.
V086881_050118

TABLE OF CONTENTS

ABOUT THE AUTHOR

James Aweya was a Senior Systems Architect with the global Telecom company Nortel, Ottawa, Canada, from 1996 to 2009. His work with Nortel involved many areas, including the design of communication networks, protocols and algorithms, switches and routers, and other Telecom and IT equipment. He received his B.Sc. (Hons.) degree in Electrical & Electronics Engineering from the University of Science & Technology, Kumasi, Ghana; M.Sc. in Electrical Engineering from the University of Saskatchewan, Saskatoon, Canada; and Ph.D. in Electrical Engineering from the University of Ottawa, Canada. He has authored more than 54 international journal papers, 39 conference papers, 43 technical reports, and received 63 U.S. patents, with more patents pending. He was awarded the 2007 Nortel Technology Award of Excellence (TAE) for his pioneering and innovative research on Timing and Synchronization across Packet and TDM Networks. He was also recognized in 2007 as one of Nortel's top 15 innovators. Dr. Aweya is a Senior Member of the IEEE. He is presently a Chief Research Scientist at EBTIC (Etisalat British Telecom Innovation Center) in Abu Dhabi, UAE, responsible for research in next-generation mobile network architectures, timing and synchronization over packet networks, indoor localization over WiFi networks, cloud RANs, software-defined networks, network function virtualization, and other areas of networking of interest to EBTIC stakeholders and partners.

PREFACE

This book discusses the design of multilayer switches, sometimes called switch/ routers, starting with the basic concepts and then on to the basic architectures. It describes the evolution of multilayer switch designs and highlights the major performance issues affecting each design. The need to build faster multilayer switches has been addressed over the years in a variety of ways and the book discusses these in various chapters. In particular, we examine the architectural constraints imposed by the various multilayer switch designs. The design issues discussed include performance, implementation complexity, and scalability to higher speeds.

The goal of the book is not to present an exhaustive list or taxonomy of design alternatives but to use strategically selected designs (some of which are a bit old, but still represent contemporary designs) to highlight the design philosophy behind each design. The selection of the example designs does not in any way suggest a preference for one vendor design or product over the other. The selection is based purely on how representative a design covers the topics of interest and also on how much information (available in the public domain) could be gathered on the particular design to enable a proper coverage of the topics. The designs selected tend to be representative of the majority of the other designs not discussed in the book.

Even today, most designs still adopt the old approaches highlighted in the book. A design itself might have existed for some time, but the design concepts have stayed pretty much alive in the telecommunication (Telecoms) industry as common practice. The functions and features of the multilayer switch have stayed very much the same over the years. What has mainly changed is the use of advances in higher density manufacturing technologies, high-speed electronics, faster processors, lower power consumption architectures, and optical technologies to achieve higher forwarding speeds, improved device scalability and reliability, and reduced implementation complexity.

As emphasized above, these design examples are representative enough to cover the relevant concepts and ideas needed to understand how multilayer switches are designed and deployed today. The book is purposely written in a simple style and language to allow readers to easily understand and appreciate the material presented.

James Aweya
Etisalat British Telecom Innovation Center (EBTIC)
Khalifa University
Abu Dhabi, UAE

1

INTRODUCTION TO SWITCH/ ROUTER ARCHITECTURES

1.1 INTRODUCING THE MULTILAYER SWITCH

The term multilayer switch (or equivalently switch/router) in this book refers to a networking device that performs both Open Systems Interconnection (OSI) network reference model Layer 2 and Layer 3 forwarding of packets (Figure 1.1). The Layer 3 forwarding functions are typically based on the Internet Protocol (IP), while the Layer 2 functions are based on Ethernet. The Layer 2 forwarding function is responsible for forwarding packets (Ethernet frames) within a Layer 2 broadcast domain or Virtual Local Area Network (VLAN). The Layer 3 forwarding function is responsible for forwarding an IP packet from one subnetwork, network or VLAN to an another subnetwork, network, or VLAN.

The IP subnetwork could be created based on well-known IP subnetworking rules and guidelines or as a VLAN. A VLAN is a logical group of devices that can span one or more physically separate network segments that are configured to intercommunicate as if they were connected to one physical Layer 2 broadcast domain. Even though the devices may be located on a number of different physical or geographically separate network segments, the devices can intercommunicate as if they are all connected to one physical broadcast domain.

For the Layer 3 forwarding functions to work, the routing functions in the multilayer switch learn about other networks, paths to destination networks and destinations, through dynamic IP routing protocols or via static/manual configuration information

Switch/Router Architectures: Shared-Bus and Shared-Memory Based Systems, First Edition. James Aweya.
© 2018 The Institute of Electrical and Electronics Engineers, Inc. Published 2018 by John Wiley & Sons, Inc.

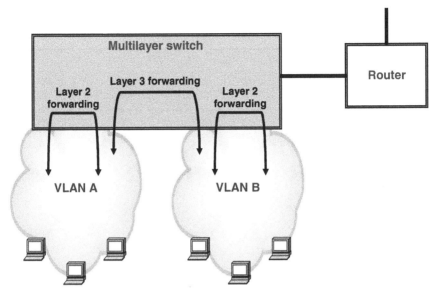

FIGURE 1.1   Layer 2 forwarding versus Layer 3 forwarding.

provided by a network administrator. The dynamic IP routing protocols – RIP (Routing Information Protocol), OSPF (Open Shortest Path First) Protocol, IS–IS (Intermediate System-to-Intermediate System) Protocol, BGP (Border Gateway Protocol) – allow routers and switch/routers to communicate and distribute network topology information between themselves and provide updates when the network topology changes occur. The routers and switch/routers via the routing protocols learn about the network topology to try to select the best loop-free path on which to forward a packet from its source to its destination IP address.

1.1.1   Control and Data Planes in the Multilayer Switch

The Layer 3 and Layer 2 forwarding functions can each be split into subfunctions – the control plane and data (or forwarding) plane functions (Figure 1.2). Comprehensive discussion of the basic architectures of routers is given in [AWEYA2000] and [AWEYA2001]. The Layer 2 functions in an Ethernet switch and switch/router involve relatively very simple control and data plane operations.

The data plane operations in Layer 2 switches involve MAC address learning (to discover the ports on which new addresses are located), frame flooding (for frames with unknown addresses), frame filtering, and frame forwarding (using a MAC address table showing MAC address to port mappings). The corresponding control plane operations in the Layer 2 devices involve running network loop prevention protocols such as the various variants of the Spanning Tree Protocol (STP), link aggregation-related protocols, device management and configuration tools, and so on.

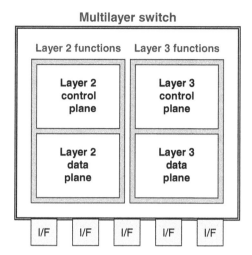

I/F = Interface

FIGURE 1.2 Control and data planes in a multilayer switch.

Even though the Layer 2 functions can be split into two planes of control and data operations, this separation (of control plane and data plane) is usually applied to the Layer 3 functions performed by routers and switch/routers. In a router or switch/router, the entity that performs the control plane operations is referred to as the routing engine, route processor, or control engine (Figure 1.3).

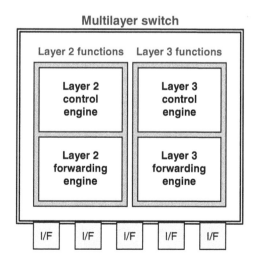

I/F = Interface

FIGURE 1.3 Control and forwarding engines in multilayer switches.

The entity that performs the data (or forwarding) plane operations is referred to as the forwarding engine or forwarding processor. By separating the control plane operations from the packet forwarding operations, a designer can effectively identify processing bottlenecks in the device. This knowledge allows the designer to develop and/or use specialized software or hardware components and processors to eliminate these bottlenecks.

1.1.2 Control Engine

Control plane operations in the router or switch/router are performed by the routing engine or route processor, which runs the operating system software that has modules that include the routing protocols, system monitoring functions, system configuration and management tools and interfaces, network traffic engineering functions, traffic management policy tools, and so on.

The control engine runs the routing protocols that maintain the routing tables from which the Layer 3 forwarding table is generated to be used by the Layer 3 forwarding engine in the router or switch/router (Figure 1.4). In addition to running other protocols such as PIM (Protocol Independent Multicast), IGMP (Internet Group Management Protocol), ICMP (Internet Control Messaging Protocol), ARP (Address Resolution Protocol), BFD (Bidirectional Forwarding Detection), and LACP (Link Aggregation Control Protocol), the control engine is responsible for maintaining sessions and exchanging protocol information with other router or network devices.

The control engine typically is the module that provides the control and monitoring functions for the entire router or switch/router, including controlling system power supplies, monitoring and controlling system temperature (via cooling fans), and monitoring system status (power supplies, cooling fans, line cards, ports

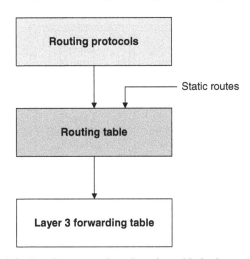

FIGURE 1.4   Routing protocols and routing table in the control engine.

Multilayer switch

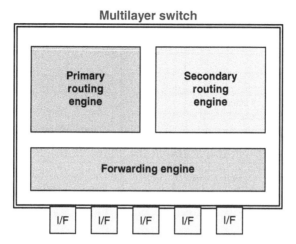

FIGURE 1.5   Multilayer switch with primary and secondary routing engines.

and interfaces, primary/secondary router processors, primary/secondary forwarding engines, etc.). The routing engine also controls the router or switch/router network management interfaces, controls some chassis components (e.g., hot-swap or OIR (online insertion and removal) status of components on the backplane), and provides the interfaces for system management and user access to the device.

In high-performance platforms, more than one routing engine can be supported in the router switch/router (Figure 1.5). If two routing engines are installed, one typically functions as the primary (or master) and the other as the secondary (or backup). In this redundant routing engine configuration, if the primary routing engine fails or is removed (for maintenance/repairs) and the secondary routing engine is configured appropriately, the latter takes over as the master routing engine.

Typically, a router or switch/router supports a set of management ports (e.g., serial port, 10/100 Mb/s Ethernet ports). These ports, generally located on the routing engine module, connect the routing engine to one or more external devices (e.g., terminal, computer) on which a network administrator can issue commands from a command-line interface (CLI) to configure and manage the device. The routing engine could support one or more USB ports that can accept a USB memory device that allows for the loading of the operating system and other system software.

In our discussion in this book, we consider the management plane as part of the control plane – not a separate plane in its own right (Figure 1.6). The management plane is considered a subplane that supports the functions used to manage the router or switch/router via some connections to external management devices (a terminal or computer). Examples of protocols supported in the management plane include Simple Network Management Protocol (SNMP), Telnet, File Transfer Protocol (FTP), Secure FTP, and Secure Shell (SSH). These management protocols allow configuring, managing, and monitoring the device as well as CLI access to the device.

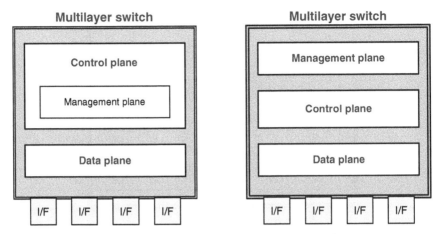

FIGURE 1.6   Control plane versus management plane.

A console port (which is an EIA/TIA-232 asynchronous serial port) could allow the connection of the routing engine to a device with a serial interface (terminal, modem, computer, etc.) through a serial cable with an RJ-45 connector (Figure 1.7). An AUX (or auxiliary) port could allow the connection of the routing engine

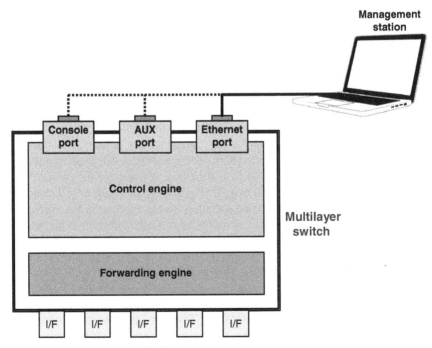

FIGURE 1.7   Management ports.

(through a serial cable with an RJ-45 connector) to a computer, modem, or other auxiliary device. Furthermore, a 10/100 Mb/s Ethernet interface could connect the routing engine to a management LAN (or a device that has an Ethernet connection) for out-of-band management of the router or switch/router.

The routing table (also called the Routing Information Base (RIB)) maintains information about the network topology around the router or switch/router and is constructed and maintained from information obtained from the dynamic routing protocols, and static routes configured by the network administrator. The routing table contains a list of routes to particular IP network destinations (or IP address prefixes). Each route is associated with a metric that is a "distance" measure used by a routing protocol in performing the best path computation to a destination.

The best path to a destination is determined by a routing protocol based on metric (quantitative value) it uses to "measure" the distance it takes to reach a destination. Different routing protocols use different metrics to measure the distance to a given destination. Then the best path to a destination selected by a routing protocol is the path with the lowest metric. Usually the routing protocol selects the best path by evaluating all the possible multiple paths available to the same destination and selects the shortest or optimum path to reach that network. Whenever multiple paths from the router to the same destination exist, each path uses a different output or egress interface on the router to reach that destination.

Typically, routing protocols have their own metrics and rules that they use to construct and update routing tables. The routing protocol generates a metric for each path through the network where the metrics may be based on a single characteristic of a path (e.g., RIP uses a hop count) or several characteristics of a path (e.g., EIGRP uses bandwidth, traffic load, delay, reliability). Some routing protocols may base route selection on multiple metrics, where they combine them into a single representative metric.

If multiple routing protocols (e.g., RIP, EIGRP, OSPF) provide a router or switch/router with different routes to the same destination, the administrative distance (AD) is used to select the preferred (or more trusted) route to that destination (Figure 1.8). The preference is given to the route that has the lowest administrative distance. The administrative distance assigned to a route generated by a particular routing protocol is a numerical value used to rank the multiple paths leading to the same destination. It is a mechanism for a router to rate the trustworthiness of a routing information source (including static routes). The administrative distance represents the trustworthiness the router places on the route. The lower the administrative distance, the more trustworthy the routing information source.

For example, considering OSPF and RIP, routes supplied by OSPF have a lower administrative distance than routes supplied by the RIP. It is not unusual for a router or switch/router to be configured with multiple routing protocols in addition to static routes. In this case, the routing table will have more than one routing information source for the same destination. For example, if the router runs both RIP and EIGRP, both routing protocols may compute different best paths to the same destination. However, RIP determines its path based on hop count, while EIGRP's

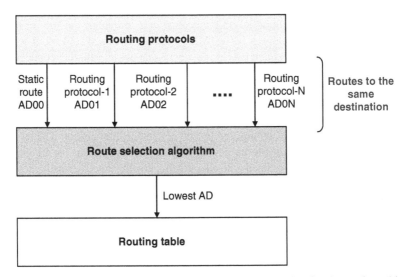

FIGURE 1.8   Use of administrative distance in route selection for the routing table.

best path is based on its composite metric. The administrative distance is used by the router to determine the route to install into its routing table. A static route takes precedence over an EIGRP discovered route, which in turn takes precedence over a RIP discovered route.

As another example, if OSPF computes a best path to a specific destination, the router first checks if an entry for that destination exists in the routing table. If no entry exists, the OSPF discovered route is installed into the routing table. If a route already exists, the router decides whether to install the OSPF discovered route based on the administrative distance of the OSPF generated route and the administrative distance of the existing route in the routing table. If the OSPF discovered route has the lowest administrative distance to the destination (compared to the route in the routing table), it is installed in the routing table. If the OSPF discovered route is not the route with the best administrative distance, it is rejected.

A routing protocol may also identify/discover multiple paths (a bundle of routes not just one) to a particular destination as the best path (Figure 1.9). This happens when the routing table has two or more paths with identical metrics to the same destination address. When the router has discovered two or more paths to a particular destination with equal cost metrics, the router can take advantage of this to forward packets to that destination over the multiple paths equally.

In the above situation, the routing table may support multiple entries where the router installs the maximum number of multiple paths allowed per destination address. The routing table will contain the single destination address, but will associate it with multiple exit router interfaces, one interface entry for each equal cost path. The router will then forward packets to that destination address across the multiple exit interfaces listed in the routing table. This feature is known as

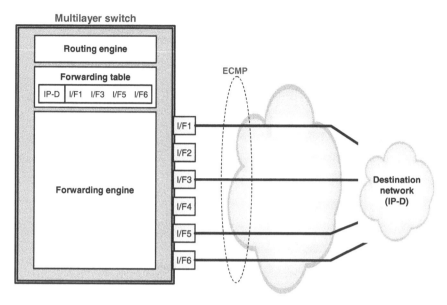

FIGURE 1.9 Equal cost multipath routing.

equal-cost multipath (ECMP) and can be employed in a router to provide load balancing or sharing across the multiple paths.

A router or switch/router may also support an important feature called virtual routing and forwarding (VRF), which is a technology that allows the router or switch/ router to support concurrently multiple independent virtual routing and forwarding table instances (Figure 1.10). VRF is a feature that can be used to create logical segmentation between different networks on the same routing platform. The routing instances are independent, thereby allowing VRF to use overlapping IP addresses even on a single interface (i.e., using subinterfaces) without conflicting with each other.

VRF allows, for example, a network path between two devices to be segmented into several independent logical paths without having to use multiple devices for each path. With VRF, the traffic paths through the routing platform are isolated, leading to increased network security, which can even eliminate the need for encryption and authentication for network traffic.

A service provider may use VRF to create separate virtual private networks (VPNs) on a single platform for its customers. For this reason, VRF is sometimes referred to as VPN routing and forwarding. Similar to VLAN-based networking where IEEE 802.1Q trunks can be used to extend a VLAN between switching domains, VRF-based networking can use IEEE 802.1Q trunks, Multiprotocol Label Switching (MPLS) tags, or Generic Routing Encapsulation (GRE) tunnels to extend and connect a path of VRFs together.

While VRF has some similarities to a logical router, which may support many routing tables, a (single) VRF instance supports only one routing table. Furthermore,

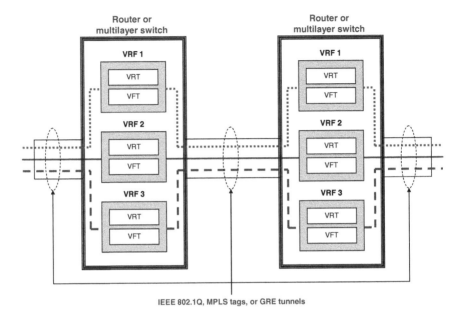

FIGURE 1.10   Virtual routing and forwarding (VRF).

VRF uses a forwarding table that specifies the next hop node for each packet forwarded, a list of nodes along a path that may forward the packet, and routing protocols and a set of forwarding rules that specify how to forward the packet. These requirements isolate traffic and prevent packets in a specific VRF from being forwarded outside its VRF path, and also prevent traffic from outside (the VRF path) from entering the specific VRF path.

The routing engine also maintains an adjacency table, which typically in its simplest form may be an ARP cache. The adjacency table (also known as Adjacency Information Base (AIB)) contains the MAC addresses and egress interfaces of all directly connected next hops and directly connected destinations (Figure 1.11). This table is populated with MAC address discoveries obtained from ARP, statically configured MAC addresses, and other Layer 2 protocol tables (e.g., Frame Relay and ATM map tables). The network administrator can explicitly configure MAC address information in the adjacency table, for example, for directly attached data servers.

The adjacency table is built from information obtained from ARP that is used by IP hosts to dynamically learn the MAC address of other IP hosts on the same Layer 2 broadcast domain (VLAN or subnet) when the target host's IP address is known. For example, an IP host that needs to know the MAC address of another IP host connected to same VLAN can send an ARP request using a broadcast address. The sending host then waits for an ARP reply from the target IP host. When received, the ARP reply includes the required MAC address and the associated IP address. This MAC address can then be used to address Ethernet frames (destination MAC address) originating from the sending IP host to the target IP host on the same VLAN.

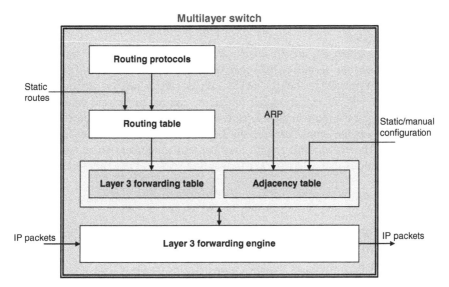

FIGURE 1.11   Layer 3 forwarding table and adjacency table.

1.1.3   Forwarding Engine

The data or forwarding plane operations (i.e., the actual forwarding of data) in the router or switch/router are performed by the forwarding engine, which can consist of software and/or hardware (ASICs) processing elements. The Layer 3 forwarding engine performs route lookup for each arriving IP packet using a Layer 3 forwarding table. In some implementations, the adjacency table is not a separate module but integrated in the Layer 3 forwarding table to allow for one lookup for all next hop forwarding information. The forwarding engine performs filtering and forwarding of incoming packets, directing outbound packets to the appropriate egress interface or interfaces (for multicast traffic) for transmission to the external network.

As already discussed, routers and switch/routers determine best paths to network destinations by sharing information about the network topology and conditions with neighboring routers. The router or switch/router communicates with their neighboring Layer 3 peers to build a comprehensive routing database (the routing table) that enables the forwarding engine to forward packets across optimum paths through the network. The information in the routing table (which is very extensive) is distilled into the much smaller Layer 3 forwarding table that is optimized for IP data plane operations.

Not all the information in the routing table is directly used or is relevant to data plane operations. The Layer 3 forwarding table (also called Forwarding Information Base (FIB)) maintains a mirror image of all the most relevant forwarding information contained in the routing table (next hop IP address, egress port(s), next hop MAC address (adjacency information)). When routing or topology changes occur in the network, the IP routing table is updated, and those changes are reflected in the forwarding table.

The forwarding engine processes packets to obtain next hop information, applies quality of service (QoS) and security filters, and implements traffic management policies (policing, packet discard, scheduling, etc.), routing policies (e.g., ECMP, VRFs), and other functions required to forward packets to the next hop along the route to their final destinations. In the switch/router, separate forwarding engines could provide the Layer 2 and Layer 3 packet forwarding functions. The forwarding engine may need to implement some special data plane operations that affect packet forwarding such as QoS and access control lists (ACLs).

Each forwarding engine can consist of the following components:

• Software and/or hardware processing module or engine, which provides the route (best path) lookup function in a forwarding table.
• Switch fabric interface modules, which use the results of the forwarding table lookup to guide and manage the transfer of packet data units across the switch fabric to the outbound interface(s). The switch interface module will be responsible for prepending internal routing tags to processed packets. The internal routing tag would typically carry information about the destination port, priority queuing, packet address rewrite, packet priority rewrite, and so on.
• Layer 2/Layer 3 processing modules, which perform Layer 2 and Layer 3 packet decapsulation and encapsulation and manage the segmentation and reassembly of packets within the router or switch/router.
• Queuing and buffer memory processing modules, which manage the buffering of (possibly, segmented) data units in the memory as well as any priority queuing requirements.

As discussed above, the forwarding table is constructed from the routing table and the ARP cache maintained by the routing engine. When an IP packet is received by the forwarding engine, a lookup is performed in the forwarding table (and adjacency table) for the next hop destination address and the appropriate outbound port, and the packet is sent to the outbound port. The Layer 3 forwarding information and ARP information can be implemented logically as one table or maintained as separate tables that can be jointly consulted when forwarding packets.

The router also decrements the IP TTL field and recomputes the IP header checksum. The router rewrites the destination MAC address of the packet with the next hop router's MAC address, and also rewrites the source MAC address with the MAC address of the outgoing Layer 3 interface. The router then recomputes the Ethernet frame checksum and finally delivers the packet out the outbound on its way to the next hop.

1.2 EVOLUTION OF MULTILAYER SWITCH ARCHITECTURES

The network devices that drive service provider networks, enterprise networks, residential networks, and the Internet have evolved architecturally and considerably over the years and are still evolving to keep up with new service requirements and user traffic. The continuous demand for more network bandwidth in addition to the introduction of new generation of services and applications are placing tremendous performance demands on networks.

Streaming and real-time audio and video, videoconferencing, online gaming, real-time business transactions, telecommuting, the increasingly sophisticated home devices and appliances, and the ubiquity of bandwidth hungry mobile devices are some of the many applications and services that are driving the need for scalability, reliability, high bandwidth, and improved quality of services in networks. As a result, network operators including the residential network owners are demanding the following features and requirements from their networks:

- The ability to cost-effectively scale their networks with minimal downtime and impact on network operations as traffic grows.
- The ability to harness the higher link bandwidths provided by the latest wireless and fiber-optic technologies to allow for transporting large volumes of traffic.
- The ability to implement mechanisms for minimizing data loss, data transfer latency, and latency variations (sometimes referred to as network jitter), thus enabling the improved support of delay-sensitive applications. This includes the ability to create differentiated services by prioritizing traffic based on application and user requirements.

The pressures of these demands have created the need for sophisticated new equipment designs for the network from the access to the core. New switch, router, and switch/router designs have emerged and continue to do so to meet the corresponding technical and performance challenges being faced in today's networks. These designs also give operators of enterprise networks and service provider networks the ability to quickly improve and scale their networks to bring new services to market.

The first generation of routers and switch/routers were designed to have the control plane and packet forwarding function share centralized processing resources resulting in poor device and, consequently, network performance as network traffic grow. In these designs, all processing functions (regardless of the offered load) must contend for a centralized, single, and finite pool of processing resources.

To handle the growing network traffic loads and at the same time harness the high-speed wireless and ultrafast fiber-optic interfaces (10, 40, and 100 Gb/s speeds), the typical high-end router and switch/router now support distributed forwarding architectures. These designs provide high forwarding capacities, largely by distributing the packet forwarding functions across modular line cards on the system. These distributed architectures enable operators to scale their networks capacity even from the platform level, that is, within a single router or switch/router chassis as network traffic loads increase without a corresponding drain on central processing resources. These distributed forwarding architectures avoid the packet forwarding throughput degradation and bottlenecks normally associated with the centralized processor forwarding architectures.

In the next chapter and also in rest of the book, we describe the various bus- and shared-memory-based forwarding architectures, starting from the architectures with centralized forwarding to those with distributed forwarding. The inner workings of these architectures are discussed in addition to their performance limitations.

1.2.1 Centralized Forwarding versus Distributed Forwarding Architectures

From a packet forwarding perspective, we can categorize broadly the switch, router, or switch/router architectures as centralized or distributed. There are architectures that fall in between these two, but focusing on these two here helps to shed light on how the designs have evolved to be what they are today.

In a centralized forwarding architecture, a processor is equipped with a forwarding function (engine) that allows it to make all the packet forwarding decisions in the system. In this architecture, the routing engine and forwarding engine both could be on one processor or the routing engine implemented on a separate centralized processor. In the simplest form of the centralized architecture, typically a single general-purpose CPU manages all control and data plane operations.

In such a centralized forwarding architecture, when a packet is received on an ingress interface or port, it is forwarded to the centralized forwarding engine. The forwarding engine examines the packet's destination address to determine the egress port on which the packet should be sent out. The forwarding engine completes all its processing and forwards the packet to the egress port to be sent to the next hop.

In the centralized architecture, there is not great distinction between a port or a line card – both have very limited packet processing capabilities from a packet forwarding perspective. A line card can have more than one port, where the line card only supports breakouts for the multiple ports on the card. Any processing and memory at a port or line card is only for receive operations from the network and data transfer to the centralized processor, and data transfer from the centralized processor and transmit to the network.

In distributed forwarding architecture, the line cards are equipped with forwarding engines that allow them to make packet forwarding decisions locally without consulting the route processor or another forwarding engine in the system. The route processor is only responsible for generating the master forwarding table that is then distributed to the line cards. It is also responsible for synchronizing the distributed forwarding tables in the line card with the master forwarding table in the route processor whenever changes occur in the master table. The updates are triggered by route and network topology changes that are captured by the routing protocols.

In a distributed forwarding architecture, when a packet is received on an ingress line card, it is sent to the local forwarding engine on the card. The local forwarding engine performs a forwarding table lookup to determine if the outbound interface is local or is on another line card in the system. If the interface is local, it forwards the packet out that local interface. If the outbound interface is located on a different line card, the packet is sent across the switch fabric directly to the egress line card, bypassing the route processor.

As will be discussed in the next chapter, some packets needing special handling (special or exception packets) will still have to be forwarded to the route processor by the line card. By offloading the packet forwarding operations to the line cards, packet forwarding performance is greatly improved in the distributed forwarding architectures.

It is important to note that in the distributed architecture, routing protocols and most other control and management protocols always run on a routing engine that is typically in one centralized (control) processor. Some distributed architectures offload the processing of other control plane protocols such as ARP, BFD, and ICMP to a line card CPU. This allows each line to handle any ARP, BFD, and ICMP messages locally without having to rely on the centralized route processor.

Current practice in distributed router and switch/router design today is to make route processing a centralized function (which also has the added benefit of supporting route processor redundancy, if required (Figure 1.5)). The control plane requires a lot of complex operations and algorithms (routing protocols, control and management protocols, system configuration and management, etc.) and so having a single place where all route processing and routing table information maintenance are done for each platform significantly reduces system complexity. Furthermore, the control plane operations tend to have a system-wide impact and also change very slowly compared to the data plane operations.

Even though the control plane is centralized, there is no need to scale route processing resources in direct proportion to the speed of the line cards being supported or added to the system in order to maintain system throughput. This is because, unlike the forwarding engine (whether centralized or distributed), the route processor performs no functions on a packet-by-packet basis. Rather, it communicates with other Layer 3 network entities and updates routing tables, and its operations can be decoupled from the packet-by-packet forwarding process.

Packet forwarding relies only on using the best-path information precomputed by the route processor. A forwarding engine performs forwarding function by consulting the forwarding table, which is a summary of the main forwarding information in routing table created by all the routing protocols as described in Figure 1.8. Based on destination address information in the IP packet header, the forwarding engine consults the forwarding table to select the appropriate output interface and forwards the packet to that interface or interfaces (for multicast traffic).

2

UNDERSTANDING SHARED-BUS AND SHARED-MEMORY SWITCH FABRICS

2.1 INTRODUCTION

One of the major components that define the performance and capabilities of a switch, switch/router, router, and almost all network devices is the switch fabric. The switch fabric (both shared or distributed) in a network device influences in a great way the following:

1. The scalability of the device and the network
2. The nonblocking characteristics of the network
3. The throughput offered to end users
4. The quality of service (QoS) offered to users

The type of buffering employed in the switch fabric and its location also play a major role in the aforementioned issues. A switch fabric, in the sense of a network device, refers to a structure that is used to interconnect multiple components or modules in a system to allow them to exchange/transfer information, sometimes, simultaneously.

Packets are transferred across the switch fabric from input ports to output ports, and sometimes, held in small temporary "queues" within the fabric when contention with other traffic prevents a packet from being delivered immediately to its destination. The switch fabric in a switch/router or router is responsible for transferring

Switch/Router Architectures: Shared-Bus and Shared-Memory Based Systems, First Edition. James Aweya.
© 2018 The Institute of Electrical and Electronics Engineers, Inc. Published 2018 by John Wiley & Sons, Inc.

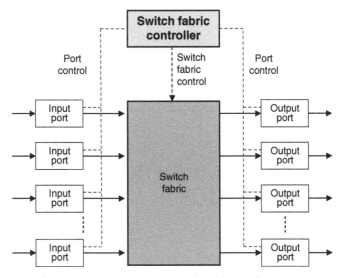

FIGURE 2.1   Generic switch fabric with input/output ports.

packets between the various functional modules (network interface cards, memory blocks, route/control processors, forwarding engines, etc.). In particular, it transports user packets transiting the device from the input modules to the appropriate output modules. Figure 2.1 illustrates the generic architecture of a switch fabric.

There exist many types of standard and user-defined switch fabric architectures, and deciding on what type of architecture to use for a particular network device usually depends on where the device will be deployed in the network and the amount and type of traffic it will be required to carry. In practice, switch fabric implementations are often a combination of basic or standard well-known architectures. Switch fabrics can generally be implemented as

- time-division switch fabrics
  - shared media
  - shared memory
- space-division switch fabrics
  - crossbar
  - multistage constructions

Time-division switch fabrics in turn can be implemented as

- shared media
  - bus architectures
  - ring architectures
- shared memory

The switch fabric is one of the most critical components in a high-performance network device and plays an important role in defining very much the switching and forwarding characteristics of the system. Under heavy network traffic load, and depending on the design, the internal switch fabric paths/channels can easily become the bottleneck, thereby limiting the overall throughput of a switch/router or router operating at the access layer or the core (backbone) of a network.

The design of the switch fabric is often complicated by other requirements such as multicasting and broadcasting, scalability, fault tolerance, and preservation of service guarantees for end-user applications (e.g., data loss, latency, and latency variation requirements). To preserve end-user latency requirements, for instance, a switch fabric may use a combination of fabric speed-up and intelligent scheduling mechanisms to guarantee predictable delays to packets sent over the fabric.

Switch/router and router implementations generally employ variations or various combinations of the basic fabric architectures: shared bus, shared memory, distributed output buffered, and crossbar switch. Most of the multistage switch fabric architectures are combinations of these basic architectures.

Switch fabric design is a very well-studied area, especially in the context of asynchronous transfer mode (ATM) switches [AHMA89,TOBA90]. In this chapter, we discuss the most common switch fabrics used in switch/router and router architectures. There are many different methods and trade-offs involved in implementing a switch fabric and its associated queuing mechanisms, and each approach has very different implications for the overall design. This chapter is not intended to be a review of all possible approaches, but presents only examples of the most common methods that are used.

2.2 SWITCH FABRIC DESIGN FUNDAMENTALS

The primary function of the shared switch fabric is to transfer data between the various modules in the device. To perform this primary function, the other functions described in Figure 2.2 are required. Switch fabric functions can be broadly separated into control path and data path functionality as shown in Figure 2.2. The control path functions include data path scheduling (e.g., node interconnectivity, memory allocation), control parameter setting for the data path (e.g., class of service, time of service), and flow and congestion control (e.g., flow control signals, backpressure mechanisms, packet discard). The data path functions include input to output data transfer and buffering. Buffering is an essential element for the proper operation of any switch fabric and is needed to absorb traffic when there are any mismatches between the input line rates and the output line service rates.

In an output buffered switch, packets traversing the switch are stored in output buffers at their destination output ports. The use of multiple separate queues at each output port isolates packet flows to the port queues from each other and reduces packet loss due to contention at the output port when it is oversubscribed. With this,

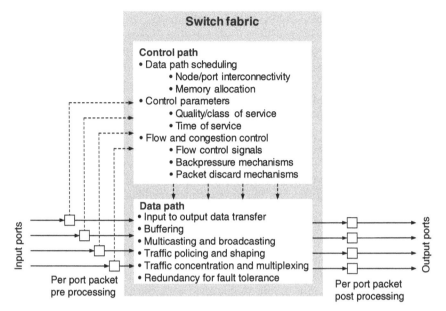

FIGURE 2.2   Functions and partitioning of functions in a switch fabric.

when port oversubscription occurs, the separate queues at the output buffered switch port constrain packet loss to only oversubscribed output queues.

By using separate queues and thereby reducing delays due to contention at the output ports, output buffered switches make it possible to control packet latency through the system, which is an important requirement for supporting QoS in a network device. The shared memory switch is one particular example of output buffered switches.

In an input buffered switch, packets are buffered at input ports as they arrive at the switch. Each input port buffering has a path into the switch fabric that runs at, at least, line speed. The switch fabric may or may not implement a fabric speed-up. Access to the switch fabric may be controlled by a fabric arbiter that resolves contention for access to the fabric itself and also to output ports. This arbiter may be required to schedule packet transfers across the fabric.

When the switch fabric runs at line speed, the memories used for the input buffering only need to run at the maximum port speed. The memory bandwidth in this case is not proportional to the number of input ports, so it is possible to implement scalable switch fabrics that can support a large number of ports with low-cost, lower speed memories.

An important issue that can severely limit the throughput of input buffered switches is head-of-line (HOL) blocking. If simple FIFO (first-in first-out) is used at each input buffer of the input buffered switch, and all input ports are loaded at 100% utilization with uniformly distributed traffic, HOL blocking can reduce the overall switch throughput to 58% of the maximum aggregate input rate [KAROLM87].

Studies have shown that HOL blocking can be eliminated by using per destination port buffering at each input port (called virtual output queues (VoQs)) and appropriate scheduling algorithms. Using specially designed input scheduling algorithms, input buffered switches with VoQs can eliminate HOL blocking entirely and achieve 100% throughput [MCKEOW96].

It is common practice in switch/router and router design to segment variable-length packets into small, fixed-sized chunks or units (cells) for transport across the switch fabric and also before writing into memory. This simplifies buffering and scheduling and makes packet transfers across the device more predictable. However, the main disadvantage to a buffer memory that uses fixed-size units is that memory usage can be inefficient when a packet is not a multiple of the unit size (slightly larger).

The last cell of a packet may not be completely filled with data when the packet is segmented into equal-size cells. For example, if a 64 bytes cell size is used, a packet of 65 bytes will require two cells (first cell of 64 bytes actual data and second cell of 1 byte actual data). This means 128 bytes of memory will be used to store the 65 bytes of actual data, resulting in about 50% efficiency of memory use.

Another disadvantage of using fixed size units is that all cells of a packet in the memory must be appropriately linked so that the cells can be reassembled to form the entire packet before further processing and transmission. The additional storage required for the information linking the cells, and the bandwidth needed to access these data can be a challenge to implement at higher speeds.

We describe below some of the typical design approaches for switch/router and router switch fabrics. Depending on the technology used, a large capacity switch fabric can be either realized with a single large switch fabric to handle the rated capacity or implemented with smaller switch fabrics as a building block. Using building blocks, a large-capacity switch can be realized by connecting a number of such blocks into a network of switch fabrics. Needless to say, endless variations of these designs can be imagined, but the example presented here are the most common fabrics found in switches/routers and routers.

2.3 TYPES OF BLOCKING IN SWITCH FABRICS

The following are the main types of data blocking in switch fabric:

- **Internal Blocking:** Internal blocking occurs in the internal paths, channels, or links of a switch fabric.
- **Output Blocking:** A switch that is internally nonblocking can be blocking at an output of a switch fabric due to conflicting requests to the port.
- **Head-of-Line Blocking:** HOL blocking can occur at input ports that have strictly FIFO queuing. Buffered packets are served in a FIFO manner. Packets not forwarded due to output conflict are buffered leading to more data transfer

delay. A packet at the front of a queue facing blocking prevents the next packet in the queue from being delivered to a noncontending output, resulting in reduced throughput of a switch.

Resolving Internal Blocking in Shared Bus and Shared Memory Architectures: Internal nonblocking in these architectures can be achieved by using a high-capacity switch fabric with bandwidth equal to or greater than the aggregate capacity of the connected network interfaces.

Resolving Output Blocking in Shared Bus And Shared Memory Architectures:

- Switch fabrics that do not support a scheduler for allocating/dedicating timeslots for packets (at the input interfaces) can have output port conflicts, which means output conflict resolution is needed on slot-by-slot basis.
- Output conflicts can be resolved by polling each input one at a time (e.g., round-robin scheduling, token circulation). However, this is not scalable when the system has a large number of inputs. Also, outputs without conflicts (just served) have an unfair advantage in receiving more data (getting a new transmission timeslot)

Resolving HOL Blocking in Shared Bus and Shared Memory Architectures:

- The system can allow packets behind a HOL blocked packet to contend for outputs.
- A practical solution is to implement at each input port multiple buffers called VoQs, one for each output. In this case, if the next packet cannot be transmitted due to HOL blocking, another packet from another VoQ is transmitted.

2.4 EMERGING REQUIREMENTS FOR HIGH-PERFORMANCE SWITCH FABRICS

In the early days of networking, network devices were based on shared bus switch fabric architectures. The shared bus switch fabric served its purpose well for the requirements of switches, switch/routers, routers, and other devices at that time. However, based on the demands placed on the performance of networks today, a new set of requirements has emerged for switches, switch/routers, and routers.

- **High-Throughput:** Switch fabrics are required to sustain very high link utilization under bursty and heavy traffic load conditions. Also, with the advent of 1, 10, 40, and 100 Gb/s Ethernet, network devices now demand correspondingly higher switch fabric bandwidth.

- **Wire-Speed Performance:** The switch fabrics are required to deliver true wire-speed performance on any one of their attached interfaces. For high-performance switch fabrics, the design constraints are, typically, chosen to ensure the fabric sustains wire-speed performance even under worst case network and traffic conditions. The switch fabric has to deliver full wire-speed performance even when subjected to the minimum expected packet size (without any typical packet size assumption). Also, the performance of the switch fabric has to be independent of input and output port configuration and assignments (no assumptions about the traffic locality on the switch fabric).

- **Scalability:** Switch fabrics are required to support an architecture that scales up in capacity and number of ports. As the amount of traffic in the network increases, the switch fabric must be able to scale up accordingly. The ability to accommodate more slots in a single chassis contributes to overall network scalability.

- **Modularity:** Switch/routers and routers are now required to have a modular architecture with flexibility to allow users to add or mix and match the number/type of line modules, as needed.

- **Quality of Service:** Users now depend on networks to handle different traffic types with different QoS requirements. Thus, switch fabrics will be required to provide multiple priority queuing levels to support different traffic types.

- **Multicasting:** More applications are emerging that utilize multicast transport. These applications include distribution of news, financial data, software, video, audio, and multiperson conferencing. Therefore, the percentage of multicast traffic traversing the switch fabric is increasing over time. The switch fabric is required to support efficient multicasting capabilities which, in some designs, might include hardware replication of packets.

- **High Availability:** Multigigabit and terabit switch/routers and routers are being deployed in the core of enterprise networks and the Internet. Traffic from thousands of individual users pass through the switch fabric at any given time. Thus, the robustness and overall availability of the switch fabric becomes a critical important design factor. The switch fabric must enable reliable and fault-tolerant solutions suitable for enterprise and carrier class applications.

- **Product Diversity:** Vendors now support a family of products at various price/performance points. Vendors continuously seek to deliver switch/routers and routers with differing levels of functionality, performance, and price. To control expenses while migrating networks to the service-enabled Internet, it is important that service providers have an assortment of products supporting distributed architectures and high-speed interfaces. This breadth of choice gives service providers the flexibility to install the equipment with the mix of network connection types, port capacity and density, footprint, and corresponding cost that best matches the needs of each service provider site. The switch fabric plays an important role here.

- **Low Power Consumption and Smaller Rack Space:** In addition to the challenge of designing a scalable system with guaranteed QoS, designers must build switch fabrics to consume minimal power while squeezing them into smaller and smaller amounts of rack space.

- **Hot Swap:** The ability to hot swap, that is, to replace or add line cards without interrupting system operations, is particularly important for high-end switch/ routers and routers. This capability is obviously an important contributor to the overall uptime and availability of the system.

Like most networking components, switch fabric designs involve trade-offs between performance, complexity, and cost. Today's most common switch designs vary greatly in their ability to handle multiple gigabit-level links.

2.5 SHARED BUS FABRIC

The simplest shared bus switch fabric comprises a single-signal channel medium over which all traffic between the system modules are transported. A shared bus is limited in capacity, length, and the overhead required for arbitrating access to the shared bus. The key design constraints here are the bus width (number of parallel bits placed on the bus) and speed (i.e., rate at which the bus is clocked, in MHz). The difficulty is designing a shared bus and arbitration mechanism that is fast enough to support a large number of multigigabit speed ports with nonblocking performance. Figures 2.3 and 2.4 illustrate high-level architectures of a bus-based switch fabric.

When multiple devices (e.g., network interface cards) simultaneously compete for access and control of the shared bus, arbitration is the process that determines which of the device gains access and control of the shared bus. Each device may be

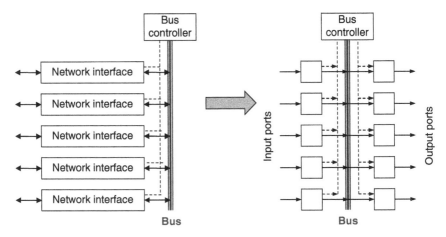

FIGURE 2.3   Shared bus fabric–single bus system.

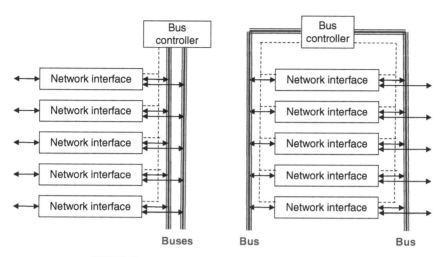

FIGURE 2.4   Shared bus fabric–multiple bus system.

assigned a priority level for bus access, which is known as an arbitration level. This can be used to determine which device should gain access and control the bus during contention for the shared bus. The switch fabric may have a fairness mechanism, which ensures that each device gets a turn to access and control the bus, even if it has a low arbitration level.

The fairness mechanism ensures that none of the devices is locked out of the shared bus and that each device can gain access to the bus within a given period of time. The central arbitration control point or the bus controller (shown in Figures 2.3 and 2.4) is the point in the system where contending devices send their arbitration signals. A simple bus implementation would use a time-division multiplexed (TDM) scheme for bus arbitration where each device is given equal access to the bus in a round-robin fashion. Because of its simplicity, the shared bus switch fabric was the most common fabric used in early routers and even in current low-end routers. The shared bus architecture presents the simplest and most cost-effective solutions for low-speed switching and routing platforms.

A big disadvantage of the shared bus switch fabric is that traffic from the slowest speed port in a shared bus system cannot speed up enough to traverse a very high-speed bus. This typically requires intermediate buffering at the slow-speed port, which further increases both the complexity and the cost of the system. In addition, issues with the hot swappability of network interface cards and fair access to bandwidth (when ports have very different speeds and traffic loads) add further complications to the design.

The typical shared bus often can be defined by the following features:

- **Bus Width:** Given that the shared bus is the signal channel/pathway over which the information from the system modules is carried, the wider the shared bus (number of bit lanes), the greater the amount of information carried over the bus. The width of the control and address buses are generally specified independent of the data bus width. The address bus width defines the number of different memory locations the bus can transfer data to.

• **Bus Speed:** The speed of the shared bus defines how many bits of data can be transferred across each lane/line of the bus in each second. A simple bus may transfer 1 bit of data per line in a single clock cycle. Some buses may transfer 2 bits of data per clock cycle, doubling performance.

• **Bus Bandwidth:** The shared bus bandwidth is the total amount of data that can be transferred across the bus in a given interval of time. If the bus width is the number of lanes and the bus speed is known, then the bus bandwidth is the product of the bus width (in bits) and the bus speed (bits per second). This defines the amount of data the bus can transfer in a second.

• **Data and Control Buses:** The typical shared bus consists of at least two distinct parts: the data bus and the control bus. The data bus consists of the signal channels/lines that actually carry the data being transferred from one module to another. The control bus consists of the signal channels/lines that carry information on how the bus functions (i.e., the control information), and how users of the bus are signaled when data are available on the data bus. An address bus may be included, which is the set of lines that carry information about where in memory the data to be transferred is stored.

• **Burst Mode:** Some shared buses can transfer data in a burst mode, where multiple sets of data can be transmitted back-to-back (sequentially in a row).

• **Bus Interfacing:** In a system that has multiple different buses, a bus interfacing circuit called a "bridge" can be used to interconnect the buses and allow devices on the different buses to communicate with each other.

The characteristics of the bus-based architecture are summarized as follows [RAATIKP04]:

• Switching over the bus is done in time domain, but implementations using both time and space switching are also possible (through the use of multiple buses).

• Bus fabrics are easy to implement and normally have low cost.

• Multicasting and broadcasting of traffic are easy to implement on the bus architecture.

• On the bus fabric only one transmission (timeslot) can be carried/propagated on the bus at any given time, which can result in limited throughput, scalability, and low number of network interfaces.

• Achieving internal nonblocking in bus architectures and implementations require a high-capacity bus with bandwidth equal to or greater than the aggregate capacity of the connected network interfaces.

• Multiple buses can be used to increase the throughput and improve the reliability of bus-based architectures.

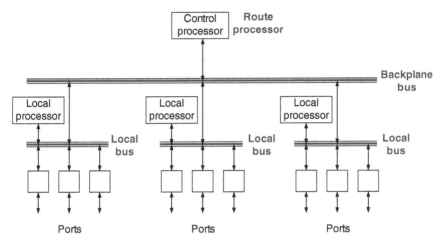

FIGURE 2.5   Hierarchical bus-based architecture.

2.6   HIERARCHICAL BUS-BASED ARCHITECTURE

Figure 2.5 shows a high-level view of hierarchical bus architecture. In this architecture, only packets traveling between local buses cross the backplane bus. In a hierarchical bus architecture, the main backplane bus is typically configured to have usable bandwidth less than the aggregate bandwidth of all the ports in the system. In such a configuration, the hierarchical bus-based switch operates well only when most of the traffic traversing the switch can be locally switched, meaning, traffic crossing the backplane bus is limited.

A major limitation of the hierarchical bus-based architecture is that when the traffic transiting the switch is not localized (to a local bus), the backplane bus can become a bottleneck, thereby limiting the overall throughput of the system. Furthermore, performing port assignments in order to localize communication to the local buses would introduce unnecessary constraints on the network topology and also make network configuration and management very difficult.

2.7   DISTRIBUTED OUTPUT BUFFERED FABRIC

Figures 2.6 and 2.7 show high-level architectures of the distributed output buffered switch. The switch fabric has separate and independent channels (buses) that interconnect any two (pairs of) input and output ports resulting in N^2 paths in total. In this architecture, packets that arrive on an input are broadcasted on separate buses (channels) that connect to each output port. Each output port has an address filter that allows it to determine which packets are destined to it.

The packets that are destined to the output are filtered by the address filters to local output queues. This architecture provides many attractive switch fabric

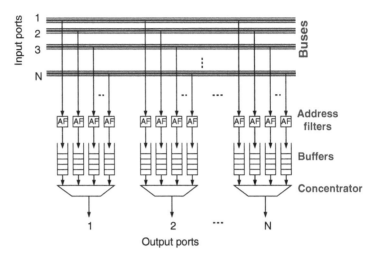

FIGURE 2.6   Distributed output buffered switch fabric: separate buffers per input port.

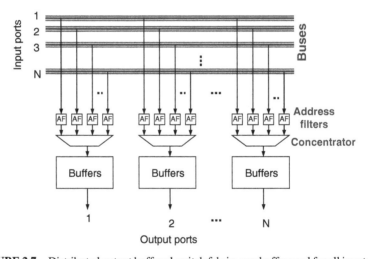

FIGURE 2.7   Distributed output buffered switch fabric: one buffer pool for all input ports.

capabilities. Obviously, no conflicts exist among the N^2 independent paths interconnecting the inputs and outputs, and all packet queuing takes place at the output ports.

Another feature is that the fabric operates in a broadcast-and-select manner, allowing it to support the forwarding of multicast and broadcast traffic inherently. Given that no conflicts exist among the paths, the fabric is strictly nonblocking and full input port bandwidth is available for traffic to any output port. With independent address filters at each port, the fabric also allows for multiple simultaneous (parallel) multicast sessions to take place without loss of fabric utilization or efficiency.

In Figure 2.6, the address filters and buffers at each port are separate and independent and need only operate at the input port speed. All of these output port components operate at the same speed. The fabric does not require any speed-up and scalability is limited to bus electronics only(operating frequency, signal propagation delay, electrical loading, etc.). For these reasons, this switch fabric architecture has been implemented in some commercial networking products. However, the N^2 growth of address filters and buffers in the fabric limits the port size N that can be implemented in a practical design.

The distributed output buffered switch fabric shown in Figure 2.7 requires a fewer number of buffers at each port; however, these output buffers must run at a speed greater than the aggregate input port speeds to avoid blocking and packet loss. The output buffer memory bandwidth and type limit the rate at which the output buffer can be accessed by the port scheduler. This factor ultimately limits the bandwidth at the output port of the switch fabric.

2.8 SHARED MEMORY SWITCH FABRIC

Figure 2.8 shows a high-level architecture of a typical shared memory fabric. This switch fabric architecture provides a pool of memory buffers that is shared among all input and output ports in the system. Typically, the fabric receives incoming packets and converts the serial bit stream to a parallel stream (over parallel lines of fixed width) that is then written sequentially into a random-access memory (RAM).

An internal routing tag (header) is typically attached/prepended to the packet before it is written into the memory. The writes and reads to the memory are governed by a system controller, which determines where in the memory the packet data are written into and retrieved from. The controller also determines the order in

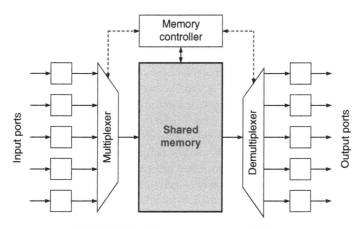

FIGURE 2.8   Shared memory switch fabric.

which packets are read out of the memory to the ports. The outgoing packet data are read from their memory locations and demultiplexed to the appropriate outputs, where they are converted from a parallel to a serial stream of bits.

A shared memory switch fabric is an output buffered switch fabric, but where the output buffers all physically reside in a common shared buffer pool. The output buffered switch has attractive features because it can achieve 100% throughput under full traffic load [KAROLM87]. A key advantage of having a common shared buffer pool is that it allows the switch fabric to minimize the total amount of buffers it should support to achieve a specified packet loss rate.

The shared buffer pool allows the switch fabric to accommodate traffic with varying dynamics and absorb large traffic bursts arriving at the system and any port. The key advantage is that a common shared buffer pool is able to take advantage of statistical sharing of the buffers as varying traffic arrives at the system. When an output port is subjected to high traffic, it can utilize more buffers until the common buffer pool is (partially or) completely filled.

Another advantage of a shared memory switch fabric is that it provides low data transfer latencies from input to output port by avoiding packet copying from port to port (only a write and read required). There is no need for copying packets from input buffers to output buffers as in other switch fabric architectures. Furthermore, the shared memory allows for the implementation of mechanisms that can be used to perform advanced queue and traffic management functions (priority queuing, priority discard, differentiated traffic scheduling, etc.).

Mechanisms can be implemented that can be used to sort incoming packets on the fly into multiple priority queues. The mechanism may include capabilities for advanced priority queuing and output scheduling. For output scheduling, the switch fabric may implement policies that determine which queues and packets get serviced at the output port. The shared memory architecture can also be implemented where shared buffers are allocated on a per port or per flow basis.

In addition, the switch fabric may implement dynamic buffer allocation policies and user-defined queue thresholds that can be used to manage buffer consumption among ports or flows and for priority discard of packets during traffic overload. For these reasons, the shared memory switch fabric has been very popular for the design of switches, switch/routers, and routers.

The main disadvantage of a shared memory architecture is that bandwidth scalability is limited by the memory access speed (bandwidth). The access speeds of memories have a physical limit, and this limit prevents the shared memory switch architecture from scaling to very high bandwidths and port speeds. Another factor is that the shared memory bandwidth has to be at least two times the aggregate system port speeds for all the ports to run at full line rate.

When the shared memory runs at full total line rate, all packets can be written into and read out from the memory resulting in nonblocking operation at the input ports. Depending on how memory is implemented and allocated, the total shared memory bandwidth may actually be a bit higher to accommodate the overhead that comes with storing variable packet sizes and the basic units of buffering. The shared memory switch fabric is generally suitable for a network device with a small number of high-speed ports or a large number of low-speed ports. It is challenging to design a shared memory switch fabric when the system has to carry a high number of multigigabit ports. At very high multigigabit speeds, it is very challenging to design the sophisticated controllers required to allocate memory to incoming packets, arbitrate access to the shared memory, and determine which packets will be transmitted next.

Furthermore, depending on the priority queuing, packet scheduling, and packet discard policies required in the system, the memory controller can be very complicated and expensive to implement to accommodate all the high-speed ports. The memory controller can be a potential bottleneck to system performance. The challenge is to implement the controller to be fast enough to manage the shared memory, read the packets on priority, and implement other service policies in the system.

The memory controller may be required to handle multiple priority queues in addition to packet reads for complex packet scheduling. The requirement of multicasting and broadcasting in the switch fabric further increases the complexity of the controller. To build high-performing switching and routing devices, in addition to a high-bandwidth shared memory fabric, a forwarding engine (ASIC or processor) has to be implemented for packet filtering, address lookup, and forwarding operations.

These additional requirements add to the cost and complexity of the shared memory switch fabric. For core networks and as network traffic grows, the wide and faster memories and controllers required for large shared memory fabrics are generally not cost-effective. This is because as the network bandwidth grows beyond the current limits of practical memory pool implementations, a shared memory switch in the core of the network is not scalable and has to be replaced with a bigger unit. Adding a redundant switching plane to a shared memory switch is complex and expensive.

In shared memory fabric implementations, fast memories are used to buffer data-arriving packets before they are transmitted. The shared memory may be organized in blocks/cells of 64 bytes to accommodate the minimum Ethernet frame size of 576 bits to handle ATM cells. This means an arriving packet bigger than the basic block size has to be segmented into the unit block size before storage in the shared memory.

The total shared memory bandwidth is equal to the clock speed per memory lane/line (in megahertz or megabits per second) times the number of lanes/lines (in bits) into the memory. The size of the shared memory fabric is normally determined from the bandwidth of the input and output ports and the required QoS for the transiting traffic. Dynamic allocation can be used to improve shared memory buffer utilization and guarantee that data will not be blocked or dropped as the inputs contend for memory space.

The characteristics of the shared memory-based architectures are summarized as follows [RAATIKP04]:

• The shared memory fabric supports switching in time domain, but implementations using time and space switching are also possible.
• Shared memory fabrics are generally easy to implement and have relatively low cost.
• Every timeslot is carried twice through the shared memory (one write and one read timeslot), resulting in low throughput, low number of network interfaces, and limited scalability.
• Achieving internal nonblocking in shared memory architectures and implementations requires a high-capacity shared memory with bandwidth equal to or greater than the aggregate capacity of the connected network interfaces.
• Multicasting and broadcasting of traffic are easy to implement on the shared memory architecture (using one write and multiple read operations from a single buffer location).

The shared memory switch fabric is also very effective in matching the speeds of different interfaces on a network device. However, the higher link speeds and the need to match very different speeds on input and output interfaces require the provisioning of a very big shared memory fabric and buffering.

As discussed earlier, the major problem in shared memory switches is the speed at which the memory can be accessed. One way to overcome this problem is to build memories with very wide buses that can load a large amount of data in a single memory cycle. However, shared memory techniques are usually not very effective in supporting very high network bandwidth requirements due to limitations of the access time of memory, that is, the precharge times, the effective burst size to amortize the storage overhead, and the width of the data bus.

2.8.1   Shared Memory Switch Fabric with Write and Read Controls

As illustrated in Figure 2.9, data from the inputs to the shared memory are time-division multiplexed (TDM), allowing only one input port at a time to store (write) a slice of data (cell) into the shared memory. Figure 2.10 illustrates generic memory architecture. As the memory write controller receives the cell, it decodes the destination output port information that is used to write the cell into a memory location that belongs to the output port and queue to receive the cell.

A free buffer address is taken from the free buffer address pool and used as the write address for the cell in the shared memory. In addition, the write controller links the write address to the tail of the output queue (managed by the memory read controller) belonging to the destination port. The read controller is signaled where the written cell is located in the shared memory.

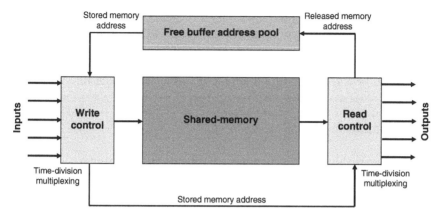

FIGURE 2.9   Shared memory switch fabric with write and read controls.

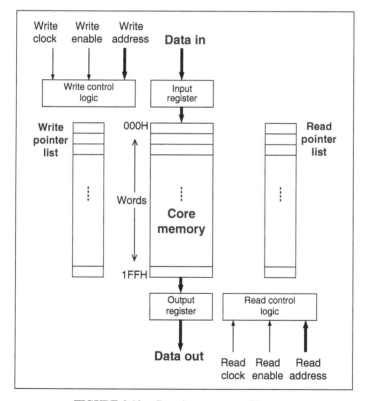

FIGURE 2.10   Generic memory architecture.

Cells transmitted out of the shared memory are time-division demultiplexed, allowing only one output port at a time to have access to (i.e., read from) the shared memory. The read process to an output port usually involves arbitration, because there may be a number of cells contending for access to the output port. The memory read controller is responsible for determining which one of contending cells wins the arbitration and is transferred to the output port. Once a cell has been forwarded to its output port, the available shared memory location is declared free and its address is returned to the free buffer address pool.

2.8.2 Generic Shared-Memory-Based Switch/Router or Router

Figures 2.11 and 2.12 both describe the architecture of a shared-memory-based switch/router or router with distributed forwarding in the line cards. Each line card has a forwarding table, an autonomous processor that functions as the distributed forwarding engine, and a small local packet memory for temporary holding of incoming packets while they are processed. A copy of the central forwarding table maintained by the route processor is propagated to the line cards to allow for local forwarding of incoming packets.

After the lookup decision is completed in a line card, the incoming packet is stored in the shared memory in buffer queues corresponding to the destination line card(s). Typically, packets are segmented into smaller size units, cells, by the line cards before storage in the shared memory. Various methods are used to signal to the

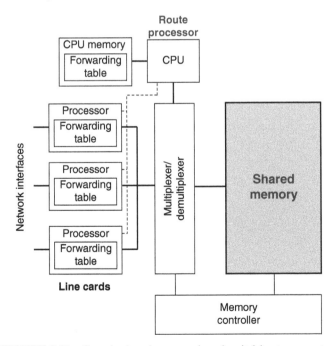

FIGURE 2.11   Generic shared-memory-based switch/router or router.

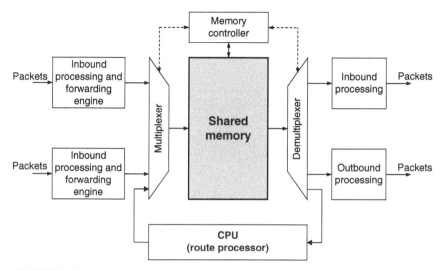

FIGURE 2.12   Generic switch/router or router with inbound and outbound processing components.

destination line card(s) that buffer queues to retrieve a processed packet. Packets stored in the shared memory can be destined to other line cards or to the router processor. A stored packet can either be a unicast packet destined to only line card or multicast, that is, destined to multiple line cards.

A broadcast packet is destined to all other line cards other than the incoming line card. Typically, the system includes mechanisms that allow for a multicast or broadcast packet to be copied multiple times from a single memory location by the destination line cards without the need to replicate the packet multiple times in other memory locations. The system also includes mechanisms that allow for priority queuing of packets at a destination port and also discarding packets when a destination port experiences overload.

The outbound processing at the destination card(s) includes IP TTL (time-to-live) update, IP checksum update, Layer 2 packet encapsulation and address rewrite, and Layer 2 checksum update. The processing could include rewriting/remarking information in the outgoing packet for QoS and security purposes.

The separate route processor is responsible for nonreal-time tasks such as running the routing protocols, sending and receiving routing protocol updates, constructing and maintaining the routing tables, and monitoring network interface status. Other tasks include monitoring system environmental status, system configuration, and line card initialization, providing network management functions (SNMP, console/Telnet/Secure Shell interface, etc.).

2.8.3   Example Shared Memory Architectures

In this section, we describe examples of shared memory switch fabrics based on Motorola's NetRAM, which is a dual-port SRAM with configurable input/output

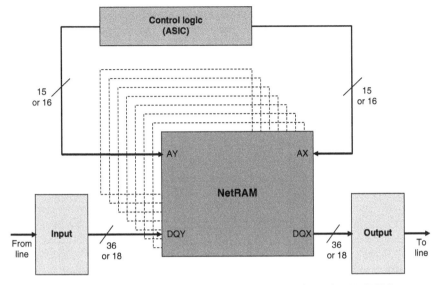

FIGURE 2.13   "One-way" switch fabric implementation using NetRAM.

data ports [MANDYLA04]. The NetRAM was designed specifically for network-ing devices that require shared memories with optimal performance for write/read/write cycles. Two implementations of NetRAM shared memory switch fabric are the "one-way" shared memory switch fabric with separate input and output ports (Figure 2.13) and the "snoop" switch fabric with two dual-ports (Figure 2.14).

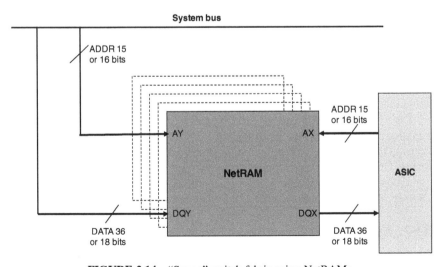

FIGURE 2.14   "Snoop" switch fabric using NetRAMs.

2.8.3.1 "One-Way" Switch Fabric Implementation Using NetRAM Figure 2.13 shows the architecture of the "one-way" shared memory switch fabric using the NetRAM. Ingress ports of line cards (or input modules) connect to the data inputs (DQY) of the NetRAM, while the egress of line cards (or output modules) connect to the data outputs (DQX). The write address ports (AY) and read address ports (AX) connect to the external control logic ASIC.

The control logic ASIC supplies the free memory address in which the cell is to be written/stored until the destination output port is ready to read/retrieve it. The free memory address is received by port AY of the NetRAM while the cell is written into the input port DQY of the NetRAM immediately after the write enable signal (WY) is activated.

When a destination output port is ready to receive a data cell, the control logic ASIC retrieves the read address of the stored cell's memory location and sends it to port AX of the NetRAM. After the NetRAM memory location is read, the data are sent out of the output port DQX after two internal clock cycles, provided the output enable (GX) signal is activated. The internal clock of the NetRAM runs at two times the external clock. This allows board frequencies to be maintained at values equal to or less than 83 MHz while the NetRAM-based device delivers data transfer performance equivalent to a conventional 166 MHz memory.

The NetRAM operates as a pipeline with reads occurring before writes. With this, if a read and a write are directed at the same memory address on the same clock edge, the cell data that were previously (written) in that memory location will be read and then the write data will be written into that location after the read has finished. This allows the NetRAM to resolve potential memory address contention problems without the additional requirement of an external arbitration mechanism or special ASICs to ensure the data transferred through the NetRAM are written and read without corruption.

Two additional functions provided by the NetRAM are a pass-through function and a write-and-pass-through function. The pass-through function allows input data to bypass the shared memory and be transferred directly from the input to an output. In situations where data need to be transferred quickly from the input port to an output port, the pass-through function can be used that saves two system cycles compared to a RAM without this function.

When data are transferred without this function from the input to the output, the output has to wait for the data to be written to the DQY port of the NetRAM (address sent through AY) and then read from the AX port (address provided through AX). The write-and-pass-through function can be used to transfer data from the input to an output, but also allowing the data to be written into the shared memory in the usual way.

The NetRAM is designed to have some advantages over conventional RAMs with single address port and common I/O port. Reference [MANDYLA04] states the main advantage to be that reads and writes can be performed at different memory addresses in the same clock cycle. This dual address read/write capability allows the NetRAM in the one-way shared memory fabric to support very high throughput of

2.98 Gb/s. In addition, implementations can be realized where several memories are banked together in parallel in the fabric. For example, if a fabric has 16 memory banks, the total available bandwidth would be 16×2.98 Gb/s (or 47.8 Gb/s).

If the data input and data output in an implementation are doubled, and reads can be performed simultaneously from both ports in the same clock cycle, then a bandwidth of approximately 6 Gb/s can be achieved. If an implementation has a 576 bit wide memory block and 16 NetRAMs connected in parallel, the maximum bandwidth becomes approximately 96 Gb/s.

2.8.3.2 "Snoop" Switch Fabric Using NetRAMs   Figure 2.14 shows a block diagram of the "snoop" shared memory switch fabric using the NetRAM. In this implementation, the NetRAM serves as a common shared memory while the system bus uses port DQX on the NetRAM for writing and reading cells. The ASIC that is connected to the dual port DQY allows the user to "snoop" into any memory address to read any data that need to be verified or modified, such as Ethernet destination MAC addresses and ATM VPI/VCI headers.

The user may want to include error checking bits or other user-defined data that the system requires as soon as the Ethernet or ATM data start to be written in the shared memory. This allows the system to make decisions at the beginning of the data input cycle, such as the next destination of the data or its service priority. Another advantage is the ability to write back data at any memory address. These features eliminate the need to have a separate system path to screen and separate out critical information from the data flow, because the entire Ethernet frame or ATM cell is stored in the shared memory. This implementation also reduces the chip count and board space. The "snoop" dual-ported switch fabric can be implemented as a $64 K \times 18$ and $32 K \times 36$ in a dual-ported device with pipelined reads.

2.8.3.3 Design Rationale of the NetRAM   For many years, burst SRAMs have been very suitable for the write-once, read-many functions of level-2 caches normally used in computing applications. To cut cost and also have access to available commercial memory devices, some designers have used burst SRAMs in the implementation of switch fabrics of network devices. Burst SRAMs generally have a common I/O port and one address port. As a result, the burst SRAM can perform either a read to an address location or write to an address location in one clock cycle but not both in the same cycle.

Also, when the common I/O on the burst SRAM has to be set from a read to write state, this requires a deselect cycle to be inserted into the timing, resulting in one wait state. The performance gets worse when turning the common I/O from a write to read state since two deselect cycles have to be inserted to ensure there are no conflicts on the burst SRAM's common I/O bus. These factors translate to a clock cycle utilization for data input or output to the burst SRAM, of between 50 and 70%, depending on the read/write and write/read patterns used to access the data in the burst SRAM.

These factors and reasons make the burst SRAM not efficient for the write/read/ write cycles of network devices. The limitations of burst SRAM become more critical when they are used for the gigabit and terabit bandwidths seen in today's switches, switch/routers, and routers. NetRAM can be implemented as a dual I/O device or a separate I/O device that eliminates the overall system performance penalties seen in the burst SRAM with a common I/O where there is the need to insert deselect cycles.

NetRAM can perform reads and writes in the same clock cycle with separate memory addresses provided for each port, allowing a designer to implement higher speed network devices than would be possible with conventional burst SRAMs. Network devices generally perform write/read/write most of the time, but the conventional burst SRAMs are suitable for the burst reads and writes of computing applications.

In network devices (switches, switch/routers, and routers), which frequently transition from reads to writes, burst SRAMs have relatively lower performance, due to the dead bus cycles that occur between operations. Thus, the use of the conventional burst SRAM increases design complexity and reduces the overall system performance. The dual port feature of the NetRAM enables the switch fabric in the network device to perform simultaneously read and write to different memory addresses in each clock cycle.

2.9 SHARED RING FABRIC

The ring fabric connects each node in the system to the immediate two adjacent nodes to it, forming a single continuous path for data traffic (Figure 2.15). Data placed on the ring travel (unidirectionally) from one node to another, allowing each

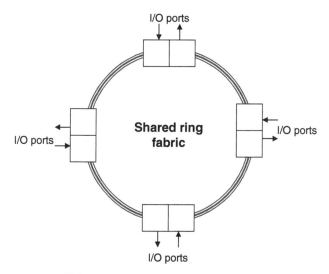

FIGURE 2.15   Shared ring switch fabric.

node along the path to retrieve and process the data. Every packet placed on the ring is visible to the nodes on the ring. Data flow in one direction on the ring hop-by-hop with each node receiving the data and then transmitting them to the next node in the ring.

The ring fabric uses controllers in the nodes that attach to the ring to manage the multiple bus segments they connect to. This fabric does not require a central controller to manage how the individual nodes access the ring. Some ring architectures support dual-counter-rotating rings to which the controllers connect. The ring fabric consists of a network of bus segments interconnected in a circular fashion.

The fabric is highly scalable because of its efficient use of internode pathways (bus segments). Adding a node to the system requires only two fabric interfaces to connect to the ring. In some ring architectures, due to the point-to-point interconnection of the nodes, it is relatively easy to install a new node since adding the node requires only inserting the node between just two existing connections. The point-to-point connectivity between nodes allows faults on the ring to be easily identified and isolated.

The ring fabric is highly scalable, although it is susceptible to single point of failure and traffic congestion. The ring is susceptible to single point of failures because it supports only one path between any two adjacent nodes. One malfunctioning node or link on the ring can disrupt communication on the entire fabric. Furthermore, adding and removing a node on the ring can disrupt communication on the fabric. Using dual (e.g., bidirectional rings) or more rings improves the reliability of the ring fabric. Generally, bidirectional ring-based architectures allow very fast reconfiguration of faults, and transmitted traffic does not require rerouting.

The bandwidth of the total ring fabric is limited by the bandwidth of the slowest link segment in the system. Furthermore, data transfer latency can be high if there are a high number of hops it takes to communicate between any two nodes in the system. Data transfer delay is directly proportional to number of nodes on the ring. This fabric architecture is advantageous where economy of system nodes (to which network interfaces are attached) is critical, but availability and throughput are less critical. Most practical systems use dual rings and identical link segments between nodes.

The characteristics of the ring-based architectures are summarized as follows [RAATIKP04]:

- Ring fabrics can generally be categorized into source and destination release rings:
  - In source release rings, only one switching operation takes place on the ring at a time, which results in limited system throughput (similar to the shared bus).
  - In destination release rings, multiple timeslots (messages) can be carried on the ring simultaneously, thus allowing spatial reuse of ring resources. This improves the throughput of the system.

- The ring fabric allows switching in time domain, but implementations using time and space switching are also possible.

- Service on the ring fabric is very orderly where every node attached to the ring has an opportunity to access (e.g., through a circulating token) and transmit data on it. The ring fabric has better performance than a shared bus fabric under heavy traffic load because of its orderly service capabilities.

- Ring fabrics are generally easy to implement and have relatively low cost (similar to shared bus architectures).

- Ring-based architectures have better scalability than shared bus ones.

- Multicasting and broadcasting of traffic are easy to implement on the ring architecture.

- Achieving internal nonblocking in ring architectures and implementations require a high-capacity ring with bandwidth equal to or greater than the aggregate capacity of the connected network interface.

- The capacity of the ring architectures can be improved by implementing parallel (multiple) rings. These rings are usually controlled in a distributed manner but Medium Access Control (MAC) implementation on the multiple rings can be difficult.

- Multiple rings can be used to reduce internal blocking, increase the throughput, and improve the scalability and reliability of ring-based architectures.

2.10 ELECTRONIC DESIGN PROBLEMS

The major problems encountered during the design of switch fabrics are summarized here [RAATIKP04]:

- **Signal Skew:** This happens when a signal placed on a line arrives at different components on the line at different times. This happens in the switch fabric and/or on circuit boards and is caused by long signal traces/lines with varying capacitive load.

- **Varying Delay on Bus Lines:** This happens when different/separate lines of a bus (with nonuniform capacitive loads) are routed through the switch fabric.

- **Crosstalk:** This happens when a signal transmitted on one line in the switch fabric creates an undesired effect on another line (i.e., electromagnetic coupling of signals from adjacent signal lines).

- **Power Supply Feeds and Voltage Swings:** When the power source/lines are incorrectly dimensioned, this can cause nonuniform voltage along a signal line. The lack of adequate filtering can also cause voltage fluctuation on the line.

- **Mismatching Timing Signals:** Lines with different lengths from a single timing source can cause phase shift in the signal received. Also, these distributed timing signals may make it difficult to have adequate synchronization.

- **Mismatching Line Termination:** This happens when terminating varying (high) bit rates inputs on long signal lines in the fabric.

Other design challenges are as follows:

- The speed of commercially available components may not necessary fit/meet the requirements of a particular design/platform (required line speeds, memory bandwidth, etc.).
- The component packing density can lead to board/circuit space utilization issues. Other problems may include system power consumption, heating, and cooling problems.
- Another challenge is balancing the maximum practical system fan-out versus the required size of the switch fabric.
- The required bus length inside a switch fabric are constrained by the following:
  - Long buses tend to require decreasing the internal speed (clock) of the switch fabric (to prevent excessive signal skew, etc.).
  - Internal switch fabric diagnostics gets difficult to carry out.

Specifically, the bus architecture itself places limits on the following:

- The operating frequency that can be used.
- Signal propagation delay that is tolerable.
- Electrical loading on the signal lines.

Propagation delay limits the physical length of a bus (distance the signal can travel without significantly degrading), while electrical loading limits the number of devices that can be connected to the bus. Pragmatically speaking, these factors are dictated by physics and cannot be easily circumvented.

3

SHARED-BUS AND SHARED-MEMORY-BASED SWITCH/ROUTER ARCHITECTURES

3.1 ARCHITECTURES WITH BUS-BASED SWITCH FABRICS AND CENTRALIZED FORWARDING ENGINES

The first generation of router and switch/router designs has relied upon centralized processing and shared memory for forwarding and buffering packets, respectively. These designs have traditionally been based on a shared bus switch fabric. In such designs, all packets received from all interfaces are written to a common memory pool.

After forwarding decisions are made, packets are then read from this shared memory and sent to the appropriate output interface(s). Even a majority of today's low-end or small-capacity systems (residential and small enterprise routers and switch/routers) still adopt this design. The simplicity and requirements of these systems make the shared-bus, shared-processor, and shared-memory architecture a natural fit.

Examples of this category of switch/router architectures are listed below. These example architectures are still in common use today and very much contemporary even though some were developed more than a decade ago. The goal here is to use these designs to highlight the main architectural approaches adopted and concepts developed over the years.

Switch/Router Architectures: Shared-Bus and Shared-Memory Based Systems, First Edition. James Aweya.
© 2018 The Institute of Electrical and Electronics Engineers, Inc. Published 2018 by John Wiley & Sons, Inc.

Example Architectures

- DECNIS 500/600 Multiprotocol Bridge/Router (Chapter 5)
- Fore Systems PowerHub multilayer switches with Simple Network Interface Modules (Chapter 6)
- Cisco Catalyst 6000 Series Switch Architectures (Chapter 7)
- Cisco Catalyst 6500 Series switches with Supervisor Engine 32 – Architectures with "Classic" line cards (Chapter 9)
- Cisco Catalyst 6500 Series switches with Supervisor Engine 32 – Architectures with CEF256 fabric-enabled line cards (optional Distributed Forwarding Card (DFC) not installed) (Chapter 9)

With the advent of high-speed, low-power electronics, multicore processor modules, and high-speed backplane technologies, these older designs have been improved to deliver even higher forwarding speeds. However, the high interface speeds and throughputs required for aggregation and core networks are driving the need for higher capacity distributed forwarding architectures that can replace the traditional system designs where line cards have to contend for a single, shared pool of memory and processing resources.

The first attempt at improving the bus-based architectures was to use multiple processor modules equipped with their own forwarding engines and resources – a pool of parallel forwarding engines. We give here a basic discussion of the major bus-based architectures that have evolved over the years.

3.1.1 Traditional Bus-Based Architecture with Software-Based Routing and Forwarding Engines in a Centralized Processor

In this architecture, a general-purpose CPU is responsible for both control plane and data plane operations (Figure 3.1). Here, an operating system process running on the general-purpose CPU is responsible for processing incoming packets and forwarding them to the correct outbound interface(s). This software-based forwarding approach is significantly slower than forwarding performed in a centralized hardware engine. The speed of the shared bus also plays a big role in the overall packet forwarding performance (throughput) of the device.

The CPU must examine the destination IP address of each packet and make forwarding decisions, rewrite MAC addresses, and forward the packet. The packet is first received and placed in a shared system memory. The forwarding process or engine (running in the centralized processor) consults the forwarding table (that may include adjacency information from the ARP cache) to determine the next hop router's IP address, outgoing interface, and the next hop's receiving interface MAC address.

The CPU then rewrites both the next hop's MAC address as the destination MAC address of the Ethernet frame carrying the packet and the MAC address of the outgoing

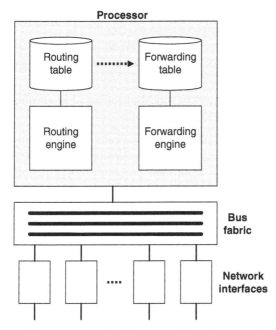

FIGURE 3.1 Traditional bus-based architecture with software routing and forwarding engines in a centralized processor.

interface as the source MAC address of the Ethernet frame, decrements the IP Time-to-Live (TTL) field, recomputes the IP header checksum and the Ethernet checksum, and finally transmits the packet out of the outgoing interface to the next hop.

This software-based forwarding architecture is slow and resource-intensive since the CPU has to process every packet in addition to performing control plane operations. For these reasons, better architectures and forwarding methods are clearly needed in order to reduce demands on the single CPU as well as increase system performance. Improvements in software-based forwarding could be achieved via the following methods (which are described in greater detail in the following sections):

- Use multiple forwarding engines in separate processors, each dedicated to packet forwarding not control plane processing.
- Add a flow/route cache to the centralized processor to enable simpler lookup operations than the more complex longest matching prefix lookup.
- Use newer highly efficient and optimized lookup algorithms with their related forwarding data structures (forwarding tables) [AWEYA2001, AWEYA2000]

In the second approach, the faster packet forwarding (using the flow/route cache) begins only after the CPU performs the slower software-based longest matching prefix forwarding on the first packet of a flow. During the software-based forwarding process, a flow/route cache entry is created that contains all the necessary forwarding information discovered during the software-based process. The flow/route cache entry allows subsequent packets of the same flow to be processed and forwarded using the faster flow/route cache instead of via the software-based full forwarding table lookup.

3.1.2   Bus-Based Architecture with Routing and Forwarding Engines in Separate Processors

Since the software-based longest matching prefix lookup in a forwarding table is processing-intensive and can be time-consuming, some software-based routers or switch/routers dedicate a separate processor solely for packet forwarding (Figure 3.2). In this architecture, one processor serves as the route processor and another as the forwarding processor (where the forwarding engine is housed).

The route processor runs the network operating system, provides all the routing functions, and also monitors and manages the components of the system. Many additional features and functions can be supported in the route processor along with the pure routing functions:

- Generating and distributing forwarding information to the forwarding processor, running control and management protocols, supporting management and configuration tools, and so on.

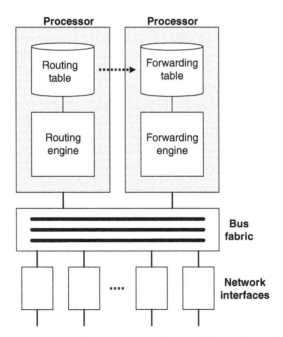

FIGURE 3.2   Bus-based architecture with routing and forwarding engines in separate processors.

- Provide out-of-band management functions and interfaces such as console, auxiliary, Ethernet, and USB ports for system configuration and management.

The forwarding processor communicates with the route processor via the shared bus or over a dedicated link or channel between them. The shared bus or dedicated link allows the route processor to transfer forwarding information to the forwarding table in the forwarding processor.

The forwarding processor can also use the dedicated link to transfer routing protocol messages and other control and management packets (ARP, ICMP, IGMP, etc.) from the external network destined for the route processor that have been received by the forwarding processor and cannot be forwarded. Packets that cannot be forwarded by the forwarding processor are considered exception packets and are sent to the route processor for further processing.

The forwarding processor provides the following functions:

- Performs forwarding table lookups using the forwarding table it maintains.
- Manages the shared memory and allocates buffers to incoming packets.
- Transfers outgoing packets to the destination ports when the forwarding decisions are made and the packets are ready to be transmitted.
- Transfers exception and control packets to the route processor for processing. Any errors originating in the forwarding processor and detected by it may be sent to the route processor using system log messages.

In this architecture, the performance of the device also depends heavily on the speed of the shared bus and the forwarding processor.

3.1.3 Bus-Based Architecture with Forwarding Using a Flow/Route Cache in Centralized Processor

In this architecture, the CPU maintains a flow/route cache that holds recently used forwarding table entries as a front-end table for forwarding packets. The entries in the cache are structured in a fast and efficient lookup format, which are consulted before the main (full) forwarding table (Figures 3.3 and 3.4). When a new packet arrives, this flow/route cache is consulted before the forwarding table maintained by the CPU.

The flow/route cache provides a simpler and faster front-end lookup mechanism that requires less processing than the main forwarding table that requires longest prefix matching lookups (in software). If the software forwarding process (engine) finds a matching entry for a destination during the flow/route cache lookup, it will forward the packet immediately to the destination port(s) and not bother consulting the forwarding table.

The relatively more extensive and complex forwarding table is only consulted when there is no entry in the flow/route cache for an arriving packet (first packet of a

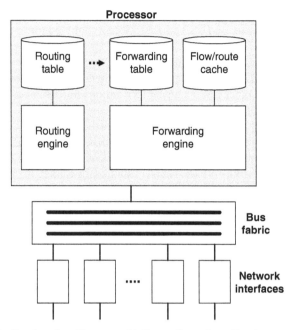

FIGURE 3.3 Bus-based architecture with forwarding using a flow/route cache in centralized processor.

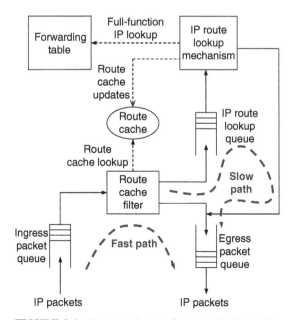

FIGURE 3.4 Route cache as a front-end lookup table.

new flow). The flow/route cache is populated with forwarding information only when the first packet has been processed. An entry in the flow/cache specifies the key information required to forward subsequent packets associated with that first packet (i.e., egress port and MAC address rewrite information that maps to a particular destination address).

The entries in the flow/route cache are generated by forwarding the first packet in software, after which the relevant forwarding information for the forwarded first packet are used to create the required information for the cache entry. Subsequent packets associated with the flow are then forwarded using the faster flow/route cache. The flow/route cache entries for a flow may include information required for QoS processing and security filtering (priority queuing, packet discard profile, packet priority value remarking, etc.).

In this flow/route cache-based architecture, some types of packets will still require extensive software processing by the CPU:

- Packets destined to the router or switch/router itself (e.g., management and control traffic, routing protocol messages, etc.)
- Packets that are too complex for the flow/route cache to handle (e.g., IP packets requiring fragmentation, packets with IP options, packets requiring encryption, NAT (Network Address Translation) traffic, etc.)
- Packets that require additional information that is not currently available or known (e.g., packets requiring destination MAC rewrites and requiring the CPU to send ARP requests)

3.1.4 Bus-Based Architecture with Forwarding Using an Optimized Lookup System in Centralized Processor

The routing table (constructed and maintained by the routing protocols) is not optimized for the packet-by-packet data plane operations. The entries in this table contain information such as the routing protocol that learned/discovered a route, metric associated with that route, and possibly the administrative distance of the route.

Although all this information is important to the overall routing process, not all is directly useable or relevant to data plane operations. This information is distilled instead to generate a smaller table, the forwarding table (or FIB), with contents more relevant for the data plane forwarding operations.

The forwarding table can be further optimized using specialized data structures to minimize data storage space while at the same time allowing faster lookups. Most often these optimized tables are tailored for specialized lookup algorithms. The forwarding process in the CPU can then use this smaller optimized lookup table and corresponding optimized lookup algorithms for packet forwarding (Figure 3.5).

This optimized lookup table organizes the forwarding information in such a way that only the routing information required for data plane operations (e.g., destination prefix, next hop, egress interface) is prominent. The optimized table may also include

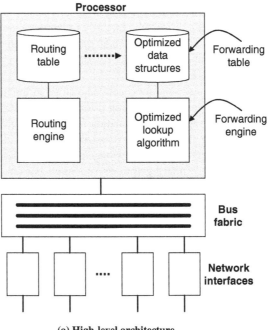

(a) High-level architecture

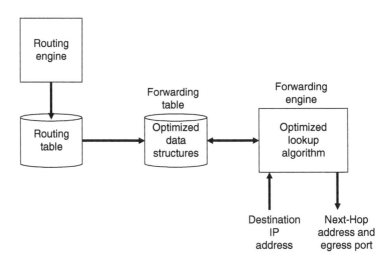

(b) Routing and forwarding components

FIGURE 3.5 Bus-based architecture with forwarding using an optimized lookup system in centralized processor.

a pointer to another optimized adjacency table, which describes the MAC address associated with the various next hop devices in the network. New forwarding table lookup algorithms have been developed over the past decade in attempts to build even faster routers. The optimized lookup tables and lookup algorithms are created with the goal of achieving very high forwarding rates [AWEYA2001, AWEYA2000]. Some architectures use optimized data structures and algorithms on specialized lookup engines/processors or ASICs engines.

More often, each lookup algorithm (which performs longest matching prefix lookup) has its corresponding optimized data structure and lookup table. These longest matching prefix lookup algorithms and their corresponding forwarding tables are typically designed as a composite structure.

3.2 ARCHITECTURES WITH BUS-BASED SWITCH FABRICS AND DISTRIBUTED FORWARDING ENGINES

Improvement in the performance of the shared-bus based architectures can be obtained by distributing the packet forwarding operations to other processors or to the line cards. Some architectures distribute forwarding engines and flow/route caches, in addition to receive and transmit buffers, to the line cards to reduce the load on the system bus and also improve overall system performance.

Other architectures distribute forwarding engines plus full forwarding tables (not flow/route caches) to the line cards to allow them to locally forward packets directly to other line cards without directly involving the route processor. Examples of the latter architectures are listed herein.

Example Architectures

- Fore Systems PowerHub multilayer switches with Intelligent Network Interface Modules (INMs) (Chapter 6)
- Cisco Catalyst 6500 Series switches with Supervisor Engine 32 – Architectures with CEF256 fabric-enabled line cards with optional Distributed Forwarding Card (DFC) installed (Chapter 9)

3.2.1 Bus-Based Architecture with Multiple Parallel Forwarding Engines

To handle traffic from a large number of interfaces or high-speed ports, some architectures employ a number of parallel forwarding engines all dedicated to packet forwarding (Figure 3.6). With a pool of multiple forwarding engines, the system can implement some form of load sharing on these engines as the traffic load increases [AWEYA2001, AWEYA2000]. Another advantage of this architecture is that it is capable of handling efficiently input interfaces with different speeds and utilization levels.

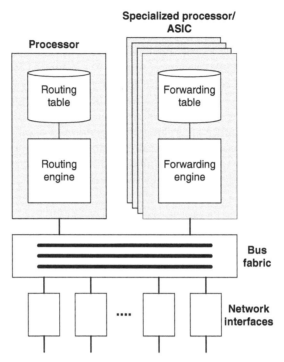

FIGURE 3.6   Bus-based architecture with multiple parallel forwarding engines.

Each forwarding engine can support specialized and optimized forwarding architectures to support not only high-speed lookups but also specialized hardware that can be used to provide QoS classification and security ACLs filtering. The ACLs can be processed at the same time when the next hop lookup is being performed.

The forwarding engines can be implemented in specialized ASICs, TCAMs, and NPUs (network processing units). ASIC implementations, in particular, allow these additional features to be turned on without affecting overall packet forwarding performance.

Using ASICs in the forwarding engines allow very high packet forwarding rates, but the trade-off is that ASICs are not very flexible and limited in their functionality because they are generally hardwired to perform specific operations and tasks. Some routers and switch/router architectures employ, instead, NPUs that are designed to overcome the inflexibility of using ASICs in the forwarding engine.

Unlike ASICs, NPUs are programmable, flexible enough to allow relatively more complex operations and features to be implemented in a forwarding engine. Also, when bug fixes and feature modifications and future feature upgrades are required, their software and firmware can be changed with relative ease.

The forwarding engines are responsible for the forwarding table lookups and packet forwarding functions in the system, while the centralized route processor provides the following functions:

- Run the routing protocols, construct and maintain the routing table, and perform all protocol message exchanges and communications with peers in the network.
- Generate the forwarding table from the routing table and distribute to the forwarding engines.
- Synchronize the contents of the multiple forwarding tables to the master forwarding table maintained by the route processor.
- Receive and process exception and control packets sent from the forwarding engines for processing.
- Load the operating system software images and other operating features (QoS and security ACL policies, priority queuing, packet discard policies, etc.) to all forwarding engines upon system power up or through operator commands.
- Provide out-of-band configuration and management of the overall system using console, auxiliary, Ethernet, and USB ports.
- Monitor and regulate the temperature of system components such as the forwarding engine modules, router processor modules, line cards, power supplies, and so on. Temperature sensors and cooling fans are used to regulate the temperature in the system.
- In a redundant configuration with a primary and secondary route processor, the system will synchronize the state of the secondary route processor to that of the primary and perform high-availability failover when the primary route processor fails.
- Serve as the central point in the system for configuring stateful firewall policies and distribution to the forwarding engines.
- Serve as the central point in the system where IP security authentication, encryption methods, and encryption keys (Internet Key Exchange (IKE)) are negotiated and maintained.

The route processor may also be the central point in the system where a wide range of IP network services, such as MPLS, Layer 2 virtual private network (L2VPN), and Layer 3 virtual private network (L3VPN) are provisioned and then configured in the forwarding engines.

The route processor in the other router and switch/router architectures is also capable of supporting the above functions. This list in fact enumerates all the potential capabilities of the route processor in a typical router or switch/router.

The forwarding engines may communicate with the route processor using a dedicated communication link. The dedicated link may be used to transfer forwarding information from the route processor to the forwarding table in the forwarding engines. Alternatively, the communication between the route processor and forwarding engines can be done through the existing switch fabric in the system.

3.2.2   Bus-Based Architecture with Forwarding Engine and Flow Cache in Line Cards

In this architecture (Figure 3.7), each line card is equipped with a flow/route cache with a relatively simple forwarding engine that allows for fast lookups in the cache. This simple forwarding engine is not as sophisticated as the full-blown forwarding engine that supports the more complex longest matching prefix lookup algorithm. As already described, the flow/route cache serves as a front-end lookup mechanism that is consulted first before the main forwarding engine in a centralized processor.

Let us assume that the first packet sent in a flow is received by a port on a line card in the switch/router. The line card examines the destination IP address in the packet and performs a lookup in the flow/route cache for an entry that is associated with this packet's destination. The line card discovers that there is no entry for this packet

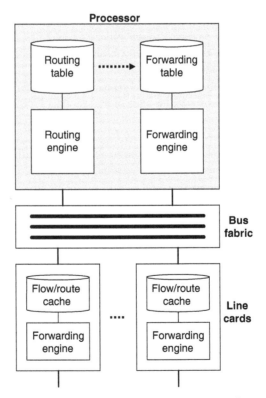

FIGURE 3.7   Bus-based architecture with forwarding engine and flow cache in line cards.

because this is the first packet of a flow, so the packet is forwarded to the (forwarding engine in the) route processor for (full) forwarding table lookup and forwarding. The line card being not able to fully process the packet may write an incomplete (or partial) flow entry in the flow/route cache at this stage of the forwarding process. This incomplete entry may include only partial information such as the source and destination IP addresses that identifies the flow the packet belongs to. By writing a partial entry at this stage, the line card will have less information to write later so that it can have spare processing cycles to handle new packets arriving.

The forwarding engine in the route processor receives the IP packet, reads the destination IP address, and performs a (longest prefix matching) lookup in its local forwarding table to determine the next hop information (next hop address, egress port, and next hop's MAC address). Typically, the lookup in the route processor is software based and only performed for the first packet of a flow.

The route processor then checks its local ARP cache (i.e., adjacency table) to determine the MAC address of the receiving interface of the next hop. If the ARP cache does not contain an entry for the next hop, the route processor transmits an ARP request (out the egress port of the next hop) for the MAC address associated with the next hop.

After obtaining the next hop's MAC address, the route processor rewrites the destination MAC address in the outgoing frame to be that of the next hop's MAC address, and the packet is forwarded out the egress port to the next hop. The source MAC address in the outgoing frame is the MAC of the outgoing interface.

After processing the first packet, the route processor sends the forwarding information associated with that packet to the line card so that it can create a complete flow/route cache entry for it. Subsequent packets in the same flow arriving at the line card do not have to be forwarded to the route processor again. The packet lookups and rewrites for these packets are performed instead in the line card using the flow/cache entry created for the first packet.

Similarly, in this architecture, some type of packets will still require handling by the route processor. These special packets have already been described.

3.2.3   Bus-Based Architecture with Fully Distributed Forwarding Engines in Line Cards

As discussed earlier, packet forwarding in the distributed forwarding architectures is done locally by distributed forwarding engines in the line cards. A master forwarding table maintained by the route processor is downloaded to the forwarding engine (ASICs or NPUs) in the line cards so that they can perform packet forwarding locally. By allowing for the forwarding to be done at the line card level (Figure 3.8), the overall forwarding throughput of the router or switch/router can be significantly increased.

Each line card is given its own memory for packet buffering and also for priority queuing of packets. A receive memory stores packets received from an interface, and a transmit memory stores packets ready for transmission out the output

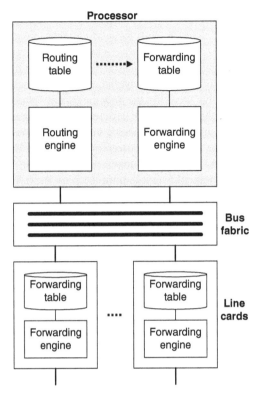

FIGURE 3.8   Bus-based architecture with fully distributed forwarding engines in line cards.

interface. Each line card also has a copy of the route processor's forwarding table and other QoS classification and security ACLs and policies so that it can forward packets without direct route processor intervention.

Some type of packets will still require handling by the route processor:

- The line card may not produce a valid path for an arriving packet that could result in the packet requiring further processing elsewhere. If the forwarding table lookup fails to find a valid entry, the packet (in many cases) is forwarded (punted) to the route processor for further processing.
- A particular feature for Layer 2, Layer 3, or QoS and security processing (MPLS, L2VPN, and L3VPN, etc.) may not be supported at the line card, thereby requiring route processor intervention.

Also, when the line card has an incomplete adjacency information for a next hop (next hop's MAC address), the line card may forward these packets to the route processor in order to start the address resolution process (ARP), which results in the adjacency information being completed some time later.

3.3 ARCHITECTURES WITH SHARED-MEMORY-BASED SWITCH FABRICS AND DISTRIBUTED FORWARDING ENGINES

One approach widely used to overcome the performance limitations of bus-based architectures is to use architectures that employ high-speed shared memory switch fabrics and distributed forwarding engines as illustrated in Figure 3.9. A key advantage of this architecture is that, in addition to serving as a switch fabric, the shared-memory can be used to temporarily store packets in large buffer memories to absorb the traffic bursts and temporary congestion that frequently occur in networks.

The memory also allows the system to queue packets using a priority queuing mechanism while they wait for transmission out the egress ports. The system can implement a number of weighted scheduling schemes (weighted round-robin (WRR), deficit round-robin (DRR), etc.) to service the priority queues. Packet discard policies such as tail drop, weighted random early detection (WRED) can

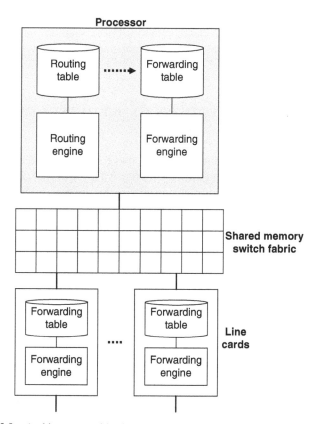

FIGURE 3.9 Architectures with shared memory-based switch fabrics and distributed forwarding engines.

also be implemented on the priority queues. The shared memory system may also support dynamic buffer allocation where buffers are allocated dynamically to the ports as traffic loads on them vary.

The shared memory switch fabric can be implemented as a single physically centralized shared memory or logically centralized shared memory (which consists of separate shared-memory modules that can be pooled together). The advantages and disadvantages of these two shared memory switch fabric designs are discussed in [CISCEVOL2001].

Example Architectures

- Cisco Catalyst 3550 Series Switches (Chapter 8)
- Cisco Catalyst 8500 campus switch routers (Chapter 10)

This distributed forwarding architecture still partitions the Layer 3 forwarding functions (which can be done in different ways, depending on the specific design), and places them in the line cards that connect to the shared memory switch fabric. The route processor still provides common services such as route processing, management and configuration, power, and temperature control to all the modules in the system.

3.4   RELATING ARCHITECTURES TO MULTILAYER SWITCH TYPES

The different router architectures discussed above certainly have different characteristics depending on the application of the device in the overall network system (Figure 3.10). The most important characteristics that are most commonly associated with an architecture type are device size, form factor, performance, reliability,

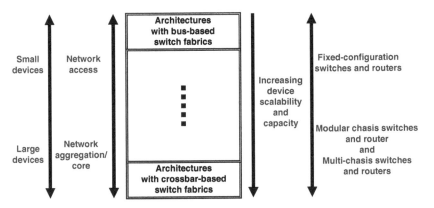

FIGURE 3.10   Relating architectures to switch types.

TABLE 3.1   Categories of Architectures Discussed in this Book

	Bus-Based Architectures	Shared-Memory-Based Architectures
Architectures with centralized forwarding engines	• DECNIS 500/600 Multiprotocol Bridge/Router • Fore Systems PowerHub multilayer switches without Intelligent Network Interface Modules (INMs) • Cisco Catalyst 6000 Series Switch Architectures • Cisco Catalyst 6500 Series switches with Supervisor Engine 32 – architectures with "classic" line cards • Cisco Catalyst 6500 Series switches with Supervisor Engine 32 – architectures with CEF256 fabric-enabled line cards (optional DFC not installed)	
Architectures with Distributed Forwarding Engines	• Fore Systems PowerHub multilayer switches with Intelligent Network Interface Modules (INMs) • Cisco Catalyst 6500 Series switches with Supervisor Engine 32 – Architectures with CEF256 fabric-enabled line cards with optional DFC installed	• Cisco Catalyst 3550 Series Switches • Cisco Catalyst 8500 campus switch routers

and scalability. Small and compact devices tend to adopt shared bus and shared memory switch fabrics, while the bigger devices are much flexible and practical to design using crossbar-based switch fabrics.

The smaller devices that are more suitable at the network access and residential networks (with a small number of user ports and lower forwarding capacities) tend to come in fixed configuration platforms. The larger devices that are usually employed at the network aggregation and core layers (and have higher forwarding capacities) tend to be based on crossbar switch fabrics and come in the form of modular chassis and multichasis platforms. The crossbar-based architecture can be designed to have advanced scalability and reliability features.

Table 3.1 presents a summary of the different types of switch/routers that are discussed in this book. The next chapters will be devoted to discussing each design in greater detail.

4

SOFTWARE REQUIREMENTS FOR SWITCH/ROUTERS

4.1 INTRODUCTION

The design and implementation of software for switch/routers and routers often requires addressing some important technical issues, namely, the processing requirements and stability of the Layer 3 (routing) and Layer 2 (bridging) protocols and algorithms, performance of the packet forwarding mechanisms, interactions between the Layer 2 and Layer 3 functions, and network management requirements. Designers of switch/routers and routers nowadays use mostly licensed or open-source (and most often enhanced) real-time kernel software and protocol software. Designers generally adopt high-quality simulation, development, and testing methods when developing product-grade networking devices.

To minimize time, effort, and cost associated with device and network management, switch/routers and routers support a combination of command-line interface (CLI) features, scripts, web-based interfaces and tools, and menu-driven configuration programs and tools, which are carefully integrated into the overall system software. The ultimate design goal of the system software and hardware is to obtain high system performance and maximized packet forwarding rate while minimizing system latency for packets.

This chapter describes the most important technical issues encountered during the development of switch/router software and the industry practices and solutions involved in the development process [COBBGR93]. This chapter describes the

Switch/Router Architectures: Shared-Bus and Shared-Memory Based Systems, First Edition. James Aweya.
© 2018 The Institute of Electrical and Electronics Engineers, Inc. Published 2018 by John Wiley & Sons, Inc.

complex nature of the design decisions involved and shows the many conflicting technical requirements that have to be addressed. While achieving extremely high forwarding rate is desirable, the design also has to address the equally important requirement of routing protocol updates and network stability (i.e., the routing process remains stable under extreme network loads). A careful balancing of these requirements is key to coming out with a high-performing yet cost-effective design.

4.2 SWITCH/ROUTER SOFTWARE DEVELOPMENT METHODS

As discussed above, the software development for a switch/router requires a real-time operating system (OS), high-performance routing, and bridging protocol software, as well as high-quality simulation and testing methods. This section describes some of the key techniques used in these system development areas.

4.2.1 Kernel Software

Currently, there is a large range of developed and refined/enhanced real-time kernel software, both licensed and open-source, to address the real-time software design constraints seen in switch/routers. An equipment vendor may choose to develop a common switch/router interface model that permits different kernels to be used to create specific platforms as required. In many cases, a vendor would use a common portable kernel that permits quick retargeting of the total switch/router software for short time-to-market development needs as well as reuse of already developed software.

4.2.2 Software Implementation

The following approaches are typically used today in the development of routing, bridging, and network management protocol software for switch/routers:

- Designer develops/implements software independently and directly from standards-based or proprietary specifications.
- Designer imports and refines/enhances software from other well-known implementations.
- Designer uses software from open-source implementations, sometimes with refinements/enhancements.
- Designer licenses software from a software supplier, for example, a software development company, government-funded university projects, and so on.

In the early years of networking, until the mid-1990s, designers developed in-house high-performance implementations of the main routing and bridging protocols (IP, IPX, Integrated IS–IS, Ethernet, Token Ring, FDDI, etc.). Designers also used specific software kernels and provided extensions for any special and proprietary networking features required. In addition, designers enhanced/extended the real-time software kernels available at that time with software interfaces commonly used in public domain kernel software (e.g., the Berkeley Software Development (BSD) UNIX socket model and system services). The addition of these software interfaces facilitated and accelerated the addition of new software from external sources.

In today's practices, a vendor would use a common switch/router and router software across its many internetworking platforms. A majority of the vendor's routing and bridging software is independent of the underlying hardware and is developed to support the evolving networking standards and also to allow portability. When designing each platform, the vendor would customize the performance-intensive and hardware-specific software code to optimize the design and maximize performance for each instance of a switch/router or router architecture.

4.2.3 Switch/Router Software Design Issues

The switch/router designer will typically have to tackle a number of technical problems related to software design when building the switch/router. This section describes the most challenging issues and how they were commonly addressed in the design process. The following are the most significant issues:

- Stability of the Layer 3 (routing) and Layer 2 (bridging) protocols and algorithms
- Requirements for network management
- Switch/router performance
- Interactions between Layer 3 (routing) and Layer 2 (bridging) functions

In addition, the total amount of memory available (system-wide) is a major constraint on the design of the switch/router. The key issues to watch out for are memory size and its usage. Typically, the memory is largely consumed by software code and by the routing and forwarding tables. These tables can be very large and are used by the switch/router to maintain the best routes to network destinations.

In the case of switch/routers that also support virtual circuits or connections (e.g., ATM, MPLS, and connection-oriented Ethernet), significant amounts of memory can be consumed by the per connection state and counter-information the device maintains. Examples of connection-oriented Ethernet technologies include IEEE 802.1ah (Provider Backbone Bridges (PBB)), IEEE 802.1Qay-2009 (Provider Backbone Bridge Traffic Engineering (PBB-TE)), and MPLS-TP (formerly T-MPLS).

Routers and switch/routers do communicate with their peers in the network to determine the best routes to a network destination. To accomplish this, all such routing devices must each be able to support the route database required to maintain a view of the network. A switch/router or router design may include an automatic shutdown mechanism that takes effect should the device run out of memory in which the routing information is stored. This mechanism, when used, can help prevent routing loops in the network.

Another design issue is the mechanisms needed to handle internal control plane and data plane congestion. To implement congestion control for traffic generated by protocols such as TCP, a router determines whether or not a packet experiences congestion by calculating the average queue length at an interface over a given time period and comparing this with predefined thresholds.

Packets may be dropped or marked for dropping later when these thresholds are breached. The calculation of the average queue values and the execution of the dropping/marking of packets must be performed in an efficient manner in real time given the potential traffic load that can converge at the device. For these reasons, and also to support real-time control, designers develop and implement algorithms specific to the particular queue structures and hardware architecture in the switch/ router or router.

4.3 STABILITY OF THE ROUTING PROTOCOLS

The stability of the routing protocols employed in the switch/router is another important issue to be considered during the design process. The system design must ensure stable and continuous operations when running the routing protocols even when the device is operating under high traffic loads.

Using a dynamic routing protocol in a network requires that routers participating in the development of the network topology map (that is used to make decisions on how packets are forwarded) must all agree on the correct path to be used at any given time. Otherwise, packets will be dropped or may loop in the network. For example, a packet will be discarded if it is sent to a router that does not know how to reach its intended destination. A packet may loop, for example, if each of the two routers believes the other is the correct next node on the path to the ultimate destination. In this case, the packet will loop between the two routers.

If a network topology never changes, and the routers and data links never get overloaded, then guaranteeing successful and continuous operations of the routing protocols would be made easier. Unfortunately, real networks are complex, dynamic, with topology changes and updates, and nodes and links can get overloaded. In real networks, the best path is agreed upon using routing protocols and algorithms distributed among multiple independent routers and operating in an ever-changing traffic environment.

The dynamic routing protocol and algorithm must converge rapidly (i.e., reach best routes agreement rapidly) so that when network conditions change, new best routes can again be agreed upon quickly. In addition, the protocol and algorithm must also be stable and not oscillate (between results).

When network traffic conditions and topology changes occur at a rapid rate, or when the routing algorithm has just completed or is trying to complete route computations, the algorithm operating among all the routers involved must still converge to a consistent state. This way, the network stays correctly operational without packet drops or routing loops.

Furthermore, while the network conditions are changing, a link or router may suddenly be introduced or be active with excessive load of packets to be forwarded (e.g., due to a transient routing loop occurring). The router must not allow this situation to upset or disturb the stability of the routing protocol and algorithm.

The stability of a well-designed and high-performing routing protocol and algorithm directly relates to how well the routing protocol and algorithm satisfies the following key requirements:

- **Link Speed between Routers:** The effective speed of the links interconnecting the routers must be high enough to allow the routers in the network to rapidly exchange routing protocol information. This aspect of routing protocol stability is under the control of the network designer, where network size and link speeds can also be continuously monitored and related to routing protocol performance.

- **Processing Power**: The route processor or control engine (control CPU) must have enough processing speed to forward routing updates to the peer routers in the network with minimum delay, and must be able to recompute the routing and forwarding databases quickly. This requirement relates only to the control CPU bandwidth available for routing protocol functions. A control CPU that devotes a lot of its processing cycles to other nonrouting protocol-related jobs will have less CPU power available, unless routing protocols jobs are given priority over these other functions. Consequently, most switch/routers and routers, especially high-end ones, now use dedicated control CPUs instead of attempting to have routing protocol tasks share the control CPU with other functions.

- **Queuing of Routing Control Messages:** The switch/router or router must ensure that end-user data and routing control messages are properly queued or separated internally so that unusually high or excessive end-user data forwarding loads do not degrade the processing of routing control messages. If not, when the router is affected by a network overload condition, the routing protocols and algorithm may not converge to a consistent state to rectify the condition. Routing stability is the most important issue that has to be properly addressed to ensure that the overall network is self-stabilizing. As the network condition worsens, routing convergence to a consistent state should not become slower. As the network condition changes more rapidly, the routing protocol and algorithm ability to recalculate the best routes must not get slower. A router that gives priority to end-user data forwarding over the reception and processing of routing control message may cause a permanent routing loop in a portion of a network and thus isolate that portion from the rest of the network.

• **Memory Usage**. The switch/router or router should be designed such that nonrouting-related functions should not consume the system memory to the point that there is not enough left over for routing control processes to carry out their function. It is important to note that even in a dedicated control CPU, some nonrouting-related activities will still take place. For example, the control CPU may have to run important activities such as network management and accounting, but which are not as critical as maintaining network stability. Obviously, without a stable network, packet forwarding, network management, and accounting will not function properly or even at all. Therefore, the design should ensure that these nonrouting-related activities do not starve the routing control processes of memory. Consequently, preallocating memory to routing control processes is the preferred practice instead of using a single critical memory pool within the router for all activities.

The next three sections describe the issues related to processing power, queuing, and memory allocation in the design of switch/routers.

4.3.1 Requirements on Processing Power

A majority of routing protocols (link-state protocols in particular) requires that routing protocol updates be received and then propagated within a specified time frame of their arrival. It is recognized that the calculations required for maintaining the routing tables are CPU-intensive, but this processing time is proportional to the number of links reported in link-state packets (LSPs), in OSPF routing, for example.

The routing update process time requirement/constraints (i.e., the time within which a routing update has to propagate at a node) mean that the CPU time must be fairly allocated between the routing decision process (i.e., time required for routing database calculations) and the routing update process itself. If the switch/router or router is required to wait until the routing decision process is completed before the routing update process is handled, then the delays on forwarding LSPs would be too large. To address these concerns, designers consider three possible solutions:

1. **Process Prioritization:** In this approach, the routing update process is given a strict priority over the routing decision process to allow the routing database to be updated as required. Here, the designer will have to find a solution as to how to synchronize access to the shared link-state database (LSDB) and at the same time allow the routing decision process to complete, if a misbehaving peer router generates LSPs at an excessive rate.

2. **Time-Slicing or Time-Sharing the CPU:** In this approach, both the routing update and routing decision processes are allowed to run simultaneously,

thus sharing the control CPU. This approach also requires synchronizing access to the LSDB.

3. **Voluntary Preemption**: In this approach, the routing decision process periodically checks if the routing update process requires execution and, if so, allows it. The checks can be configured to happen at time intervals frequent enough to satisfy the delay constraints/requirements and at times suitable/appropriate for the routing decision process to avoid the need for synchronizing access to the LSDB.

Designers typically choose the third approach to avoid synchronizing access to the LSDB for the following two reasons:

- In complex systems, synchronization issues can often create system problems that are serious and difficult to debug. An architecture that avoids these issues entirely can often lead to increased software simplicity and reliability.
- Adding synchronization mechanisms for parallel tasks can decrease total system performance, for example, because it causes excessive rescheduling of operations.

As a result, an architecture that uses voluntary preemption of processes allows a simpler yet very efficient solution that still meets the system requirements.

4.3.2 Queuing Requirements

By using appropriate queuing mechanisms, the control CPU can ensure that during high network traffic loads, routing control information will not be discarded. In a centralized CPU routing/forwarding architecture, separating the end-user data (to be directly forwarded) from routing control messages is a more logical solution to preserving the processing routing control information. However, this works well only if the CPU can process all the routing control messages without losing messages. Low-end switch/routers or routers typically do not have a control CPU that is fast enough to guarantee such routing information processing performance.

Even if a router can guarantee the timing requirements on the routing update and decision processes even under worst-case traffic loads (for OSPF routing, for example) and if that load is combined with a flood of Hello, Database Description (DBD), Link-State Request (LSR) messages, Link-State Update (LSU), and Link-State Acknowledgment (LSAck), some of these messages have to be queued for later processing or simply discarded. However, even in a large network, the worst-case scenario would be if all these nodes were to send Hello and other control messages at the same time, which is very unlikely.

With careful software design, the switch/routers or routers can meet the network stability requirements and still not lose connectivity to their peer routing devices on the network. In architectures with limited control CPU power, the current practice is to design and implement a traffic management policy that differentiates between routing protocol message types (e.g., OSPF messages: Hello, DBD, LSR, LSU, LSAck) to meet their respective processing requirements for network stability.

The traffic management parameters that control the minimum and maximum numbers of packets allowable for each (differentiated) routing control message type have to be carefully calculated based on each message's architect behavior and the network· configuration in which the router is deployed. For example, a router designed to support a given maximum number of adjacent routers in a network will influence the selection of the traffic management policy for managing the OSPF Hello message queues and packet buffers. Such traffic management mechanisms (for the control messages) should be implemented to guarantee that the minimum levels of message exchanges (per message type) are met to ensure overall system forwarding performance, network convergence, and stability.

The traffic management policy for the routing control messages can be designed to also use buffer memory pools that support priority queuing. The priority queues allow various traffic management policies to be implemented. Inbound control traffic can be placed in the priority queues according to routing message type. The queues are then serviced using a number of priority scheduling algorithms (weighted round-robin (WRR), weighted fair queuing (WFQ), etc.).

These algorithms can be configured with different weights assigned to each priority queue to ensure that all message types are adequately served, although at different rates. The actual weights can be made configurable to take into account the performance characteristics of the router and expected network configuration it would be placed in.

Other routing control traffic management policies may include the following:

- **Separate Buffer Pools Per Routing Control Message Type**: An architecture may have a separate buffer pool for each of the different control message types. The disadvantage of this approach is that, in small network configurations or ones that do not carry heavy routing control traffic, the pool of buffers may be constantly underutilized and also denying the actual data plane packet forwarding process from utilizing these buffers when it needs them.

- **Strict Priority Scheduling of Routing Control Message Queues**: Configuring strict priority scheduling for processing the different routing control message types is undesirable, because the flooding of one routing control message type could starve one or more other message types of CPU processing opportunity for a long time. In such a case, the system works better when each message type is given some processing time than to giving one type absolute priority.

4.3.3   Requirements on Memory Allocation

Network devices, in general, must have adequate buffering for storing and processing packets, but switch/routers and routers, in particular, must have sufficient buffering to handle routing control messages. Today's routing devices strive to guarantee this buffering requirement for control messages. The line cards must never run out of buffers; otherwise, routing control messages required by the route processor may be discarded depriving the routing protocols of important network information. The route processor must also have sufficient buffering for the control messages it receives so that they are properly processed.

4.4   NETWORK MANAGEMENT

In a majority of networks, networking management constitutes the biggest portion of the cost in operating the network. Most of this cost involves recruiting and maintaining trained and experienced field engineers and technicians, network managers, and operators. Minimizing these costs requires deploying network equipment (switches, switch/routers, and routers) that have the right features and capabilities that allow for easy, fast, and efficient management. The major network management issues include the following:

- **Installation and Loading of Software:** This relates to how software updates are distributed and installed on network equipment. It also includes how long the switch/router or router takes to load after electric power interruption.
- **Router/Software Configuration:** This relates to how the switch/router or router software is configured with information about changes to the network interfaces, links, network parameters, and so on. It also deals with whether the network device requires a reboot to change information it is supplied with.
- **Network Monitoring:** This relates to how the network manager gets immediate information/reports about network problems and unexpected or sudden changes. Faults in the network can cause unacceptable network performance degradation or downtime. Thus, it is important that they are detected, logged, users notified of them, and remedied quickly to keep the network running smoothly. Network monitoring also includes getting long-term reports of network traffic patterns and capacity usage to be used in network planning. To accomplish this and other related tasks, the network needs to incorporate performance monitoring capabilities that allow for long-term network performance trending.
- **System and Access Control:** This relates to how the network manager can shut down an entire switch/router or router or just an interface/line in the system. This also deals with how to provide secured access to the network devices and resources to authorized persons.
- **Problem Solving and Debugging**: This relates to the type of tools available to detect problem in the network and to investigate and correct them.

Given that skilled network management staff may not be available at all locations (particularly in small branch offices in a service provider network), in addition to some sites not being staffed at all, remote management capabilities are also essential.

4.4.1 Installation and Loading of Software

With the current proliferation of Internet services, switch/router or routers are able to update their software (software updates or new software) over the network. Also switch/router or routers are able to store their software in nonvolatile memory and so do not need to be reloaded on each boot. However, some older systems load the software each time they are booted.

The use of nonvolatile memory (e.g., flash memory) allows for fast and reliable software loading combined with backup load operation when software updates are required. The load can be from a network server using, for example, the trivial file transfer protocol (TFTP). This provides an easy way to update software when required.

4.4.2 Router/Software Configuration

The general view about routers is that they are notoriously very difficult to configure. Whether this view is justified or not depends on which network manager you talk to and which brand of router is up for discussion. Router vendors have gone to great lengths to dispel this as a myth. In this regard, vendors have been developing tools (using the latest software technologies available) to assist network managers with configuration.

Each switch/router or router comes with configuration tools: CLI based, Secure Shell (SSH), menu-driven programs, scripts, web-based tools, and so on. For example, a menu-driven program could lead the network manager through a series of forms to be filled to define the information required to configure the router (or switch/router) or to modify an existing configuration.

The tools often come with online help information to provide extra information needed for configuration and sometimes tutorials. The help information may also provide information about how steps may be retraced. Consequently, the network manager has no need to learn very lengthy, complex, and often very technical information about configuring the device.

During the design and development of the configuration tools, router vendors normally use formal human factors testing to ensure that these tools meet the expectations of the device user (network manager). The vendor uses the router's customer field testing and trials to provide additional feedback on the configuration tool's effectiveness and ease of use. Even throughout the life of the product, the vendor continues to use human interface testing to refine the quality and perform-ance of the tools.

A network manager can use an SSH client at a remote location to make a secure, encrypted connection to a SSH server in a switch/router or router. SSH supports strong encryption for authentication, and the user authentication can be via username and password, TACACS+ and RADIUS (see TACACS+ and RADIUS below). The authentication and encryption in SSH allows an SSH client to carry out a secure communication over a network that is even insecure.

4.4.3 Network Monitoring

It is now common practice for network devices including switch/routers to include capabilities for the network to detect faults, isolate them, log, and notify the network manager about the fault conditions. The network manager or any automated mechanism can then take the necessary remedial measures to correct the faults encountered in the network to keep it running effectively. Other capabilities provided in the devices allow them to monitor and measure a number of network parameters and metrics so that the performance of the network can be maintained at the desired operational level.

Switch/router or router incorporate capabilities to allow for a wide range of performance and system metrics/parameters at the device, interface, and protocol levels to be collected at regular intervals using, for example, the Simple Network Management Protocol (SNMP). With this, a Network Management System (NMS), supporting a polling engine, allows data collection from the network (switch/routers, routers, etc.). Most current NMSs have extensive network capabilities that include, at a minimum, collecting, storing, and presenting polled data.

Currently, there exist many commercial and public domain tools that are capable of collecting, storing, and presenting data from switch/router, routers, and other network equipment. Network managers can use graphical interfaces that present network maps and show the operational states of the network devices such as switch/routers and routers. Current network devices including switch/routers and routers support web-based interfaces that enable the device's performance data to be accessible from remote locations in the network. In addition to SNMP, many switch/routers and routers support other features such NetFlow, sFlow, and RMON (Remote Network MONitoring).

RMON probes and NetFlow in the switch/routers and routers provide the ability to collect traffic profiles from them. User traffic has grown significantly in networks and continues to place greater demands on network resources. Consequently, the need to collect information on user traffic and network load has become even more important. Network managers, who in the past had a limited view or knowledge of the types of traffic running in the network, can use NetFlow, for example, to profile the network traffic. User and application traffic profiling gives the network manager even a better view of the traffic in the network.

The RMON can be deployed in a network at various locations where agents (implemented in stand-alone RMON probes or embedded in network devices) pass on information to a central management station through SNMP. The latest RMON version allows remote monitoring of a network device from the Media Access Control (MAC) sublayer up to the network and even application layers. The statistics, alarms, history, and other data provided by RMON can be used by a network manager to proactively monitor and take steps to improve network availability at various protocol layers in the network.

NetFlow allows detailed traffic flows statistics to be collected from network devices, which can then be used for billing, network and device troubleshooting, and network capacity planning. NetFlow can be implemented on individual ports, links, and interfaces on a network device, allowing information on traffic passing through these points to be collected. The NetFlow data/information gathered on a network device is exported to an entity referred to as a collector. Some of the functions performed by the collector include reducing the volume of data it receives (which involves filtering and aggregation functions), hierarchical data storage, and file system management.

Some routers provide features that can be utilized for measuring response times between routing devices. One router with such a feature is capable of measuring the response time to other routing devices. SNMP traps can be configured on the source router to alert an NMS if the response times cross some predefined thresholds.

4.4.4 System and Access Control

Switch/routers and routers, as standard practice, do have mechanisms to control administrative access to them and network resources. These access control mechanisms, for example, can monitor users logins to the device and refuse access to unauthorized users. Some of the methods of controlling access to switch/routers and routers include entering local user IDs and passwords on the device, configuring Access Control Lists (ACLs), and using protocols such as TACACS+ (Terminal Access Controller Access Control System plus) or RADIUS (Remote Authentication Dial-In User Service).

Both TACACS+ and RADIUS, although different, are protocols that provide secure remote access to networks and network services based on an Authentication, Authorization, and Accounting (AAA) architecture. AAA, generally, is an architectural framework for managing access to network services and resources, enforcing access and usage policies, auditing service and resource usage, and generating information needed for billing purposes.

The process of authentication provides mechanisms for identifying a user and determining whether that user is allowed access to the network. This often involves, in the simplest case, the user supplying a username and password that are checked against a database of usernames and passwords. The switch/router or router supports mechanism for identifying a user (network manager or administrator) prior to being allowed access to the device.

The authorization process provides mechanisms to determine what type of resources, services, or activities are permitted for an authenticated user. Authorization only occurs after authentication. Authorization can be used do define the specific services, resources, or activities allowable for the (different) users. Also, in the context of configuring a switch/router or router, authorization can be used to limit the commands that are permissible to the different authenticated users. Authorization can also include a one-time authorization and can be for each service that is requested by the user.

The accounting process provides mechanisms for tracking what a user carried out and when it was done (on authorized or unauthorized resources, services, or activities). Accounting can be used for billing for service or resource usage and for performing an audit trail. Accounting can be used to gather information on user identities, start and stop times, and the commands executed on the switch/router or router. Accounting does not need to happen after authentication and authorization, it can be carried out independent of them.

Both TACACS+ and RADIUS are client/server protocols. TACACS+ runs on TCP, while RADIUS runs on either UDP or TCP (in IETF RFC 6613, RADIUS over TCP). TACACS+ provides separate AAA services, while RADIUS combines both authentication and authorization services. TACACS+ is mostly used by the network administrator for secure access to network devices such as switches, switch/routers, and routers, while RADIUS is commonly used to authenticate and login remote users to a network.

SNMP can also be used to carry out configuration changes on switch/routers and routers similar to the capabilities offered by a CLI for system configuration. However, the right security measures have to be implemented on the switch/router or router to prevent unauthorized access and change via SNMP. SNMPv3 (SNMP version 3) supports secure exchanges of management data (by authenticating and encrypting data) between a management station and network devices.

SNMPv3's encryption and authentication features ensure that data transported to a management station are secured than in SNMPv1 and SNMPv2. When SNMP is used, the standard practice is to provide the necessary security features needed to ensure that only authorized network management stations (NMPs) are allowed to perform configuration changes on switch/router or router.

An SNMP privilege level can be used to specify/limit the types of operations that a management station can perform on the switch/router or router. A Read-Only (RO) privilege level allows a management station (NMP) to only query the device's data. This does not permit the management station to issue configuration commands such as rebooting the device and shutting down network interfaces. The Read-Write (RW) privilege level is the only mode that allows such operations to be performed. SNMP ACLs can be used with SNMP privilege levels to specify which specific management stations are allowed to request management information from the switch/routers or routers.

4.4.5 Problem Solving and Debugging

Problem solving and debugging tasks in a network constitute the biggest, most time-consuming, and expensive aspects of a network manager's responsibilities.

Fortunately, there exist today many networking techniques and tools for performing extensive and very complex tasks required for testing and debugging switch/ routers, router, and other network devices.

Building on initial experience gained during the design, testing, and debugging of a switch/router (as a matter of fact, other network devices), vendor normally produces problem-solving guides for the device. These user and problem-solving guides typically provide a step-by-step description of how various problems associated with the deployment of the device can be isolated and solved.

Typically, the vendor conducts human factor testing on these guides to determine how effective they are and also investigate different forms of making this information available. The guide is normally made available to the user in the form of portable flash memory drives, hard copy (in some cases), and also online on the vendor's Web site.

Another tool for problem solving is incorporating software and hardware features into the networking device that allows it to record or copy packets that are sent to and received on its interfaces (some sort of port-mirroring and network analyzer capabilities). The copied packets can then be analyzed and displayed automatically by software routines on an external management station.

The mirroring capability functions like having a built-in analyzer on the network interface. This feature can serve as an effective diagnostic tool used by the design engineers when debugging system hardware and software development problems, and by field engineers when investigating a problem on the customer site. Switch/ router and router vendors also include diagnostic facilities such as loopback testing over the device's interfaces.

Network problem solving and debugging can also be done using SNMP, RMON, and other similar protocols. A network management platform (NMP) receives events (or notifications) from network elements (switches, switch/routers, routers, severs, etc.) in the network and processes them for the purpose of carrying out fault, configuration, performance, security, and accounting management. A standard NMP supports the following functions: network discovery, mapping the topology of network elements, event handler, collection and plotting of performance data, and browser for data management.

An NMP can perform the discovery of the network devices and then represent/ display each network device by a graphical element on the NMP's console. The graphical elements on the console typically use dissimilar colors to indicate the current working state of a switch/router, router, and other devices in the network. To carry out the needed functions, the network devices can be configured to send notifications (SNMP traps) to the NMP. When the NMP receives the SNMP traps (also stored in a log file), it changes the graphical element representing the particular network device to a different color based on the severity of the SNMP trap received.

Important tools for network troubleshooting that can be used by the network manager are Trivial File Transfer Protocol (TFTP) and system log (syslog) servers.

The key use of the TFTP server is for storing system configuration information (files) and software code/images for switches, switch/routers, routers, and other network devices. Switch/routers and routers are capable of sending system log messages to a syslog server located somewhere in the network. The syslog messages can be used by the network manager to perform troubleshooting tasks when network problems occur. The syslog messages can also be used by network field personnel to carry out root cause analysis.

The primary goal of fault management is to detect faults encountered in a network, possibly (automatically) isolate them, and then notify other entities who will be responsible for correcting them. To do this, switches, switch/routers, routers, and other devices are now capable of alerting management stations (NMPs) when a fault occurs on the network. Sending SNMP trap messages, using SNMP polling, logging messages in a syslog server, and employing RMON thresholds all form critical components of an effective fault management system. An NMP can then alert the network manager when a network fault is reported and remedial/corrective actions are required.

SNMP traps describing the environmental conditions of a switch/router or router can be sent to the NMP when an environmental threshold is exceeded:

- A voltage notification can be sent if the voltage measured at a given point in a network device is outside the normal range (at warning, critical, or shutdown stage).
- A shutdown notification can be sent if a system parameter at a given test point in a network device is reaching a critical state and the system is about to initiate a shutdown.
- A redundant supply notification can be sent if a redundant power supply in the network device fails.
- A fan notification can be sent if any one of the cooling fans in the network device's cooling fan array fails.
- A temperature notification can be sent if the temperature at a given point in the network device is outside the normal range (at warning, critical, or shutdown stage).

A switch/router or router can implement RMON alarms and events capabilities to allow it to monitor itself for the rising and falling of a system parameter/metric against thresholds (input/output packet drops, CPU utilization, buffer failures, etc.). With this, the switch/router or router will take samples of the parameter/metric at predefined time intervals and compare it against the configured thresholds. The device will send an SNMP trap to a management station if the measured value falls outside the configured thresholds. The network manager can use RMON alarms and events provided by the network devices for proactive management of the network before potential problems occur.

4.5   SWITCH/ROUTER PERFORMANCE

Today's enterprise and service provider networks rely on high-performing switches, switch/routers, and routers to carry high volumes of diverse traffic with different quality of service (QoS) requirements. For these reasons and others, network designers network managers must understand the characteristics and capabilities of the switching and routing platforms they choose. This section discusses the performance aspects of today's switch/routers and routers.

Network performance and performance metrics play a very important role in service-level agreements (SLAs). An SLA is an agreement between a service provider and a customer on the level of performance expected of network services to be delivered by the service provider's network. At the core of the SLA are a set of metrics agreed upon between the customer and the service provider. The target values agreed upon for these metrics by both parties must be measurable, meaningful, and realistic.

A number of statistics can be collected from the network and its individual devices to measure its performance level. These statistics can be used to calculate/develop the performance metrics or included directly as metrics in the SLA. Other statistics such as input interface queue drops, output interface queue drops, and error/corrupted packets are mainly useful for diagnosing network performance-related problems.

At the network device level, performance metrics can include buffer allocation and utilization levels, memory allocation, and CPU utilization. These performance metrics can be useful because the performance of some network protocols (e.g., TCP, routing protocols) is directly related to the availability of adequate buffering in the network devices. Thus, collecting device-level performance metrics and statistics is critical when optimizing the performance of these protocols.

Various performance metrics at the device, interface, and protocol levels can be collected using SNMP. Network management systems do support a polling engine that can be used for collecting parameters and variables from network devices. The typical network management system is capable of polling, collecting, storing, and presenting the received data in various format (e.g., graphical).

Various commercial solutions are also available to address the needs of performance management of service provider and enterprise networks. These systems are designed to have extensive capabilities including collecting, storing, and displaying/presenting data from network devices and servers. They generally include web-based interfaces that make it possible to access the performance data from remote locations in the network.

4.5.1   Performance Metrics

A number of common metrics have been developed for benchmarking the performance of switch//routers and routers. The following three major metrics are commonly used:

- **Throughput:** This refers to the maximum forwarding rate (e.g., packets per second) at which no offered packets are dropped by the device.

- **Packet Loss Rate:** This is the fraction of packets out of the total offered traffic load that the network device should have forwarded while operating under a constant traffic load, but were not forwarded due to lack of resources.

- **Latency (or Transit Delay):** This refers to the time interval from when the last bit of the packet reaches the input port to when the first bit of the packet is seen on the output port.

4.5.2 Packet Throughput/Forwarding Rate

Since the packet forwarding rate is the most important performance metric for a switch, switch/router, or router, designers carefully optimized both the hardware and software of the device to allow packet forwarding to occur at as much higher speed as possible. In architectures with centralized forwarding, the designers try to optimize the design such that all the forwarding on the central CPU is done with every available software and hardware assistance possible at minimal cost.

In architectures with distributed forwarding, the design is optimized such that forwarding and filtering operations are handled by line cards. Both designs would typically employ various forms of hardware assist for the performance-critical destination address lookups in the forwarding engine. This allows the design to support the necessary requirements for very high-speed packet forwarding.

On each line card in a distributed forwarding architecture, a streamlined software kernel may be developed for the local processor along with all its required software. The line card processor software kernel and modules have to be carefully designed to have the minimum number of instructions and the lowest number of execution cycles necessary to perform the high-speed filtering and forwarding operations.

On the control or route processor of the switch/router or router (which is typically centralized), the software kernel also has to be designed to be fully capable of forwarding packets on its own. However, the route processor is mainly required to provide the software processing for the nonperformance-intensive operations of the system (i.e., the processing of routing protocol updates used to build the routing database and network management messages). The partitioning of processing of received packets (between the line cards and the route processor) in the distributed architecture allow such systems (and the networks they reside in) to remain highly stable when traffic overloads occur.

In some architectures with centralized forwarding, the software-based forwarding operation has no hardware assist at all. Instead, the software-based lookup algorithms have to be optimized and implemented to meet the performance-intensive requirement of the (software) forwarding engine. In these systems, the software is highly optimized and tuned for the processor supporting the forwarding engine to allow higher performance forwarding. The amount of software code is also kept to a minimum to minimize the additional maintenance overhead associated with highly optimized and tuned software.

4.5.3   Packet Transit Latency

The packet latency (or transit delay) in switch/routers and routers also depends heavily on the hardware and software processing and storage characteristics of the device. One design factor that affects the packet latency is whether packet processing (and address lookup) starts prior to the packet being completely received or not. Another factor (mainly in Layer 2 forwarding) is whether the packet received can be sent out to the output port for transmission prior to its complete reception.

In Layer 2 forwarding (mainly), when packet reception, forwarding, and transmission occur in parallel, the architecture is referred to as cut-through. With cut-through switching, the switch delays the arriving packet only long enough to read the relevant Layer 2 packet header and make a forwarding decision (destination MAC address and, if necessary, VLAN tag and IEEE 802.1p priority field). The cut-through switch does not have to wait for the full packet serialization by the sending device before it begins packet processing.

The internal switch latency during cut-through switch is reduced because packet processing is limited to the Layer 2 header only rather than to the entire packet. The fundamental drawback of cut-through switching is that corrupted packets cannot be identified and discarded because the packets are forwarded before the FCS field is received and thus no CRC calculation can be performed. In spite of this shortcoming, cut-through switching, which was the predominant mode of switching in the 1990s, has enjoyed resurgence in recent years because of the reduction in latency it provides.

Cut-through switching is applicable for networks that do not require speed changes and are limited enough in diameter/extent to have very low packet error rates. Low error rates mean that only a negligible amount of bandwidth will be wasted on bad packets. Typically, networks that have optical fiber connectivity have lower packet corruption than those based on copper. Fortunately, many of today's networks tend to be fiber based, which have greater immunity to electrometric interference and radio-frequency interference (EM/RFI), crosstalk, and so on. Fiber-based network have relatively low attenuation and support greater distances and bandwidths.

The store-and-forward switch has to wait for the full packet to be received (i.e., serialized by the sending device) before it begins packet processing and forwarding. The switch latency for a packet is the delay between the last bit into the switch and the first bit out (LIFO) of the switch. After the output port has been determined, the switch has to reserialize the packet to send it on its way to the receiving device.

In some routing (Layer 3) device designs, the packet reception and forwarding can be done in parallel prior to a packet being completely received. However, such designs do not start transmitting the packet at the output port until the packet is completely received because Layer 3 forwarding and address lookups are relatively indeterministic compared to Layer 2 forwarding.

In routing (Layer 3) devices, the forwarding model is store-and-forward. The factors that affect the packet latency in such devices are as follows:

- **Packet Reception:** The Layer 2 and 3 headers of the packet must be completely received before processing can start.
- **Forwarding Table Lookup Operations:** This includes Layer 2 packet header verification, parsing and analyzing the Layer 3 header, performing the required Layer 3 destination address lookup, performing any required Layer 2 and 3 packet field rewrites, and queuing the packet for transmission on the destination port/interface.
- **Queuing Due to Congestion:** If the destination interface is busy when it receives a packet, it will have to be queued until the interface is ready for transmission. The internal packet latency for a device must reflect the potential latency delays due to congestion at various points in the device including the output interface. Congestion avoidance algorithms in protocols such as TCP are designed to minimize this congestion delay when overload conditions occur.
- **Packet Transmission Delay:** This delay refers to the time taken to clock/place the bits of the packet out of the interface onto the transmission medium but may also include medium access delays (particularly in media that require media contention). In such media, delays occur due to another node already using the common shared transmission medium.

Now let us look at a number of methods used by switch/routers and routers to minimize the packet transit delay. Most single processor systems minimize the packet reception and transmission delays by allowing the line interface hardware to perform the receive and transmit functions while using their direct memory access (DMA) features. In these systems, the same data path (fast-path) optimizations used to improve the packet forwarding rate are used at the same time to minimize the forwarding delay.

In architectures with distributed forwarding, the line cards generally have no DMA, and the onboard processor receives each packet byte-by-byte and parses the Layer 2 and 3 headers as soon as they are received. The data link (Layer 2 header) fields are immediately examined and decoded even before the network (Layer 3) address bytes have been received. The destination network address lookup is initiated as soon as the address fields have been received (i.e., before the packet payload or data have been received). The destination address lookup is then performed by the local forwarding engine without direct involvement from other modules in the system.

In some designs, the line cards receive packets one chunk (segment) at a time, that is, a complete packet is received in smaller segments. Typically, these line cards use, internally, small fixed-size buffers (smaller than the largest maximum transmission unit (MTU) of the system) that are linked together, as required, to store a complete packet. The line card's forwarding engine performs the forwarding table lookup as soon as the header information is available (i.e., when the first segment of the packet is received).

The advantage of the above line card architecture is that, for a large packet, the forwarding engine is able to complete a forwarding decision well before the last byte of the packet has been received. However, the disadvantage is that, until the last byte of the packet has been received by the system, it is impossible to tell whether the CRC of the packet is correct or the packet has not been corrupted.

For these reasons, the receive line card does not pass the packet to the destination line card until the CRC check has been successfully completed. This system design is still store-and-forward (not cut-through) even though the packet forwarding decision is initiated immediately after the Layer 3 header is received.

Before a packet is transmitted to the destination interface, certain modifications must be made in the packet. In IP, the time-to-live (TTL) field is updated and, in some cases, other fields like those for QoS indication may be modified, after which the IP checksum is recalculated. Layer 2 packets may need header field modification for QoS indication or conversion between untagged and tagged Ethernet frames. To speed up packet forwarding and reduce packet latency, some line cards have hardware assistance for modifying the packet header information including updating checksums and CRCs.

The above design options all attempt to reduce the packet transit delay as much as possible using different mechanisms. These mechanisms try to reduce the transit delay by controlling the time it takes to receive a packet to the time it departs the system. The designs incorporate modules that blend some aspects of store-and-forward processing with cut-through design.

Some of these architecture process packets in distributed processors (in the line cards) but transmit packets only after complete reception and validation of the packets (store-and-forward processing). This design limits the packet forwarding rate of the system, because packets that are found to be in error are discarded, but these can only be discovered if the system receives complete packets. The full cut-through designs in comparison have relatively higher packet forwarding rates, but they would propagate corrupted/error or invalid packets that eventually get discarded by the end user.

4.6　INTERACTION BETWEEN LAYER 3 (ROUTING) AND LAYER 2 (BRIDGING) FUNCTIONS IN SWITCH/ROUTERS

The relationship and interaction between the Layer 3 (routing) and Layer 2 (bridging) functions introduces some level of complication in the design of switch/routers. A received packet must be either Layer 2 forwarded (bridged) or Layer 3 forwarded (routed), but the latter process itself involves many aspects of the Layer 2 functions. To handle the combined functions on an integrated product, several designs can be used as illustrated in Figures 4.1–4.3 [COBBGR93].

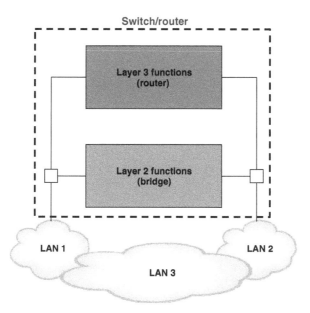

FIGURE 4.1   Integrated switch/router with protocol split. Some protocols are passed to the Layer 2 (bridging) functions, while others that are routable protocols are passed only to the Layer 3 (routing) functions.

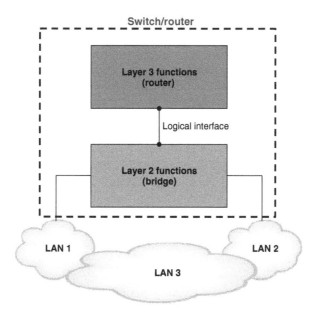

FIGURE 4.2   Integrated switch/router with shared external interface. The Layer 3 (routing) function uses its own internal Layer 2 address and logical interface to the external networks.

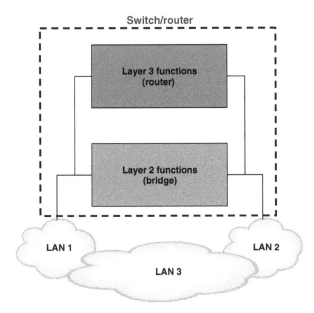

FIGURE 4.3 Integrated switch/router with multiple external interfaces. The Layer 3 (routing) function uses all Layer 2 interfaces to attach to the external network.

4.6.1   Switch/Router Design with Protocol Split

In this switch/router design approach (Figure 4.1), traffic from nonroutable proto-cols are Layer 2 forwarded (bridged), and those from routable protocols are routed, for example, IP, Internetwork Packet Exchange (IPX), Systems Network Archi-tecture (SNA), and AppleTalk. The Layer 2 and Layer 3 functions are completely separate, the only common feature is that they share the same link interfaces. Every packet received by the switch/router in this design is sent either to the Layer 3 function (if routable) or to the Layer 2 functions.

Unlike routable protocols such as IP, traffic from protocols such as DEC's Local Area Transport (LAT), NetBIOS Extended User Interface (NetBEUI) cannot be routed over a network. The nonroutable protocols do not have network (Layer 3) addresses but only have device (Layer 2) addresses. The protocols do not support any network addressing scheme that can be used to forward packets from one network to another network.

A nonroutable protocol is mainly limited and used within a LAN where all the devices belong to the same network. If the devices are located in different networks (LANs), the nonroutable protocol they use does not allow them to exchange data. They can only do that when the data they generate are encapsulated in a routable protocol.

Unless a nonroutable protocol is involved, this switch/router design approach is not usable and has no significant benefits.

4.6.2   Integrated Switch/Router Design with Shared External Interface

In this design approach (Figure 4.2), the Layer 3 function is interfaced to and layered on top of the Layer 2 function. In this case, the Layer 3 function uses its own internal Layer 2 address(es) and logical interface to interface to the Layer 2 function, which in turn connects to the external networks. Incoming packets are passed to the bridging process first, but if they are addressed to the Layer 3 function (router), they are then forwarded to the routing process. In this design, the routing function uses a logical interface within the switch/router over which it receives routable packets.

In this design, the Layer 3 and 2 functions can be logical entities within the same physical platform or they can be physically separate collocated entities. The latter architecture is similar to a Layer 2 switch connected to a "one-armed router" that provides Layer 3 functionality for inter-VLAN communications.

A one-armed router, also called a "router on a stick," is a Layer 3 forwarding device that has a one logical or physical connection to a network that has more than one VLANs (subnets) requiring routing between them. A one-armed router is often used to forward inter-VLAN traffic and connects to these VLANs via a Layer 2 switch.

4.6.3   Integrated Switch/Router with Multiple External Interfaces

This switch/router design (Figure 4.3) is similar to the design with shared external interface shown in Figure 4.2, except that the Layer 3 function uses all the interfaces also available to the Layer 2 function and logically connects to the same extended network through all these interfaces. This design constitutes the standard design found in a majority of today's switch/routers. Some older designs adopt the approach presented in Figure 4.2.

4.7   CONTROL AND MANAGEMENT OF LINE CARDS

In most distributed forwarding architectures, each line card is designed to have a separate software environment that is initialized/configured, controlled, and managed by a centralized control/route processor. Some of key tasks carried out in the line cards are as follows.

4.7.1   Watchdog Polling of the Line Cards

This feature is needed to enable the system guard against its software being stuck in an infinite loop and thereby preventing it from responding to system control and management instructions and messages. The control processor most likely will be protected by a hardware watchdog timer, but the line cards, on the other hand, may not support such a timer feature. To enable the line card software guard against such problems, the control processor software can be designed to poll each line card periodically, say, every 500 ms. With this, if the control processor receives no response, the line card is reset.

4.7.2 Statistics Counters in the Line Cards

The line cards in the distributed forwarding architecture are designed to handle packet forwarding, which in many cases creates the additional requirement of maintaining counters (e.g., number of packets in error, number of data bytes received). A designer may choose to implement these using 32 bit counters. However, to avoid implementing 64 bit counters in the line card (which may increase the line card memory costs and require 64 bit arithmetic for the counters), the designer may implement the full (bigger) counters in the centralized control processor and then allow it to poll the line cards periodically, but frequently enough to ensure that the line card counters do not overflow and wrap around. Each line card counter may be sized (i.e., number of bits) to support, for example, 400 ms polling frequency by the control processor.

4.7.3 Control of the Line Cards

The line cards can be designed such that when a Layer 2 (data link) protocol or a Layer 3 (routing) protocol is started or stopped on any one of its network interfaces, the centralized control processor would receive the necessary management commands/notifications and also allow it to issue appropriate control commands and messages to the line card.

4.8 DISTRIBUTED FORWARDING

In this architecture, each line card handles the forwarding of Layer 2 and Layer 3 packets without direct involvement of the centralized control/route processor. Designers use different approaches to ensure the switch/router meets the stability requirements of the Layer 3 (routing) functions regarding the processing of routing protocol messages, network convergence, and topology stability.

In some designs, the switch/router or router discards normal data packets to make room for the routing protocol messages in order to meet the routing stability requirements. The packet discard is implemented by appropriate traffic management mechanisms to guarantee a minimum level of memory and processing cycles for the routing messages, even when operating under worst-case network conditions such as those created by network topology changes.

Other switch/router or router designs do not need to discard normal data packets, because the line cards are designed to be able to continue forwarding normal data packets while, at the same time, the route processor handles the routing protocol operations. In addition, the line cards are designed such that they are able to ensure that routing control traffic is always passed to the route processor, even in cases where the line card is also transferring large amounts of user traffic to other line cards.

5

ARCHITECTURES WITH BUS-BASED SWITCH FABRICS: CASE STUDY— DECNIS 500/600 MULTIPROTOCOL BRIDGE/ROUTER

5.1 INTRODUCTION

The DEC Network Integration Server 500 and 600 (DECNIS 500/600), developed by Digital Equipment Corporation (also known as DEC or Digital), are examples of the earlier first-generation switch/routers that integrated (multiprotocol) Layer 3 forwarding (routing), Layer 2 forwarding (bridging), and network gateway functions on a single platform [BRYAN93, COBBGR93]. These devices offered multilayer switching over a number of LAN and WAN interface types available at that time.

The DECNIS 500/600 was developed in the early 1990s during the era of multiprotocol networking (IP, IPX, AppleTalk, FDDI, X.25, etc.) and when switches and routers were mostly implemented in software. The DECNIS 500/600 was designed to be flexible enough to support a wide range of Layer 2 and 3 functionalities while offering high forwarding performance. This chapter describes the architecture and implementation of the DECNIS 500/600 switch/route along with the main ideas that influenced the design of this family of switch/routers.

To achieve high packet forwarding performance, the DECNIS 500/600 employed a centralized forwarding engine assisted by distributed forwarding algorithms and buffer management mechanisms implemented on the line cards. This method of centralized forwarding, where only packet headers are sent to the centralized forwarding engine, and actual packet forwarding performed by the line

Switch/Router Architectures: Shared-Bus and Shared-Memory Based Systems, First Edition. James Aweya.
© 2018 The Institute of Electrical and Electronics Engineers, Inc. Published 2018 by John Wiley & Sons, Inc.

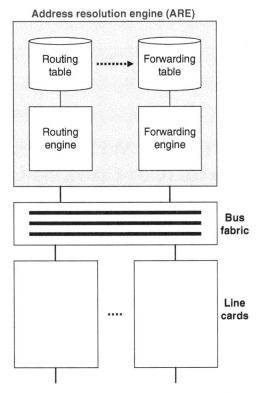

FIGURE 5.1   Bus-based architecture with forwarding engine in centralized processor.

cards is referred to as "in-place packet forwarding." Packet forwarding to the destination line card(s) is performed by the source line card assisted by memory components it supports. The discussion includes a description of the packet forwarding processes in the line cards.

Based on the architecture categories defined in Chapter 3, the architectures discussed here fall under "Architectures with Bus-Based Switch Fabrics and Centralized Forwarding Engines" (see Figure 5.1).

5.2   IN-PLACE PACKET FORWARDING IN LINE CARDS

The first generation of routers and switch/routers performed longest prefix matching lookups in software, mainly due to the complexity of the IP forwarding table lookup process. Software-based lookups, however, limited the overall packet forwarding rates of a routing device, especially when the device supports high-speed interfaces. This limitation drove routing equipment vendors to develop other techniques to improve packet forwarding even if the device uses a centralized forwarding engine.

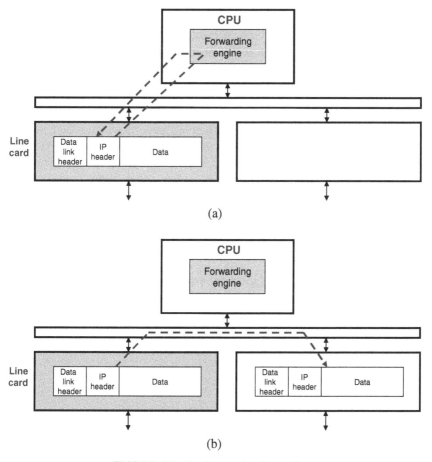

FIGURE 5.2   In-place packet forwarding.

One such forwarding method is to allow the inbound network interface (or line card) to forward only the packet header of the incoming packet to the centralized forwarding engine. The forwarding engine processes the packet header (forwarding table lookup, TTL update, IP header checksum update, Ethernet checksum update, etc.) and then instructs the inbound interface to forward the packet (along with a modified packet header) to the correct outbound interface(s) (Figure 5.2). This forwarding approach (in-place packet forwarding) improves system performance by reducing the amount of data transfers that can take place across the (bus-based) switch fabric. Data transfer over the shared bus is often the major factor that limits performance in bus-based devices).

As described in previous chapters, another approach is to allow a forwarding engine to maintain a route/flow cache of frequently used destination addresses that can be used as a front-end lookup table. This front-end table is consulted first before the main forwarding table any time a packet arrives. Several architectures have been proposed over the years that allow for exact match searches to be performed in hardware within the route/flow cache. In this case, if a destination address is matched by the route/flow search, the packet can be forwarded immediately instead of being sent for the more complex longest prefix matching lookup. These improved architectures are described in various chapters of this book.

Higher performing designs decouple the routing and management processing (control plane) from the actual packet forwarding processing (data plane). Separation of the control and forwarding planes is achieved by allowing a system to have two independent processing modules, a routing and management engine (also called the route processor) and a packet forwarding engine. With this decoupling, the control plane and data plane functions are assigned to these separate engines or processors. The forwarding engine handles the actual packet forwarding, while the route processor is responsible for running the routing protocols and generating the routing and forwarding tables used by the forwarding engine.

5.3 MAIN ARCHITECTURAL FEATURES OF THE DECNIS 500/600

The DECNIS 500/600 architecture (described in [BRYA93] and illustrated in Figure 5.3) consists of a bus-based backplane (Futurebus+) along with its

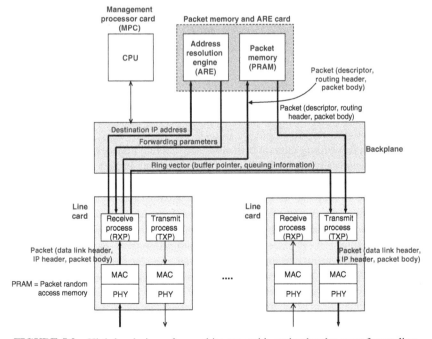

FIGURE 5.3 High-level view of an architecture with packet header-copy forwarding.

associated interface circuitry, a control CPU card (Management Process Card (MPC)) that supports the core control plane functions, a number of network interface cards (line cards), and a packet memory card (Packet Random-Access Memory (PRAM) card). The PRAM card in turn contains a centralized forwarding engine (Address Resolution Engine (ARE)) that is responsible for Layer 2 and Layer 3 forwarding of packets.

The central packet memory (PRAM) is organized in units of 256 byte buffers, and is shared among all the network interface cards. The DECNIS 600 supports seven line card slots, while the smaller DECNIS 500 supports only two line card slots. The PRAM is the module in which packets transiting or destined to the switch/router itself are stored. Buffer ownership is transferred from one network interface card to another by a swap process (where a full buffer is exchanged for an empty one).

The purpose of the buffer management mechanism is to govern and improve the performance of the transfer mechanism (handling buffer ownership), in addition to providing buffer allocation fairness. To allow efficient storage of data, the system uses fractional buffers much smaller than the maximum packet size (maximum transmission unit (MTU)) of the network interfaces in the system. The system is designed to allow applications to run from local memory supported on the line cards and the MPC. As will be discussed below, the line cards support processors to allow some packet forwarding functions to be performed locally.

The DECNIS 500/600 was designed to allow forwarding of packets directly between the line cards (once the forwarding table lookups are completed by the ARE) in order to maximize packet forwarding performance. The line cards support functions that allow Layer 3 forwarding on the line card (for a number of Layer 3 routing protocols, for example, IP and IPX) as well as Layer 2 forwarding for other Layer 2 traffic (e.g., Ethernet, FDDI). The management processor runs the software that controls the system, including initialization and start-up of the line cards, generating and exchanging routing and bridging protocol messages, construction and maintenance of the Layer 3 and Layer 2 tables, and network management.

5.4   DECNIC 500/600 FORWARDING PHILOSOPHY

When forwarding a Layer 3 (routed) packet, the incoming packet at a network interface is logically viewed to consist of three parts: the data link header, the Layer 3 header, and the packet payload (body). The receive process (RXP) shown in Figure 5.3 examines the data link layer information in the arriving packet and extracts the data link header from the packet. The RXP then parses and copies the Layer 3 header into the packet memory (PRAM) without modifying it.

Any required modifications on the Layer 3 header are carried out after the Layer 3 forwarding decision is made (by the ARE) and when the packet is ready to be transmitted. The modification information to be written in the packet is carried in a data structure called a packet descriptor (Figure 5.3). The packet descriptor occupies space reserved at the front of the allocated first packet buffer space. Recall that the

PRAM is organized in units of 256 byte buffers. The packet payload is copied into packet buffers in the PRAM, using up packet buffers as required.

After the destination IP address is copied to the ARE for forwarding table lookup, the receive process is free to start processing another incoming packet. Meanwhile, the ARE completes the forwarding table lookup process, after which the receive process reads from the ARE the forwarding parameters needed to complete the processing and forwarding of the packet. The forwarding parameters provide to the RXP information about the packet's outbound port and channel (at the output line card), the data link address for the next hop, and any packet rewrite information required for forwarding. The receive process adds the forwarding information received from the ARE to the information saved from parsing the arriving packet to construct the packet descriptor in packet memory (PRAM) (Figure 5.3).

The receive process constructs a set of ring vectors for the ARE-processed packet, one ring vector for each (256 byte) packet buffer used to store the packet. Each ring vector holds a pointer to the (256 byte) packet memory (PRAM) buffer used, in addition to other information used to determine to which priority queue the packet buffer should be assigned as well as to determine the packet's relative priority in the system. When congestion occurs, the additional ring vector information (that carries the packet's priority) is used by the network interface cards to discard the low-priority packets first.

The ring vectors associated with the processed packet are then sent to the transmit process (TXP) on the output network interface card, which then queues them (locally) to be transmitted to the next hop. To begin the process of preparing the packet for transmission, the TXP first reads the packet descriptor from its location in the first packet memory (PRAM) buffer. The information in the packet descriptor is used by the transmit process to build (rewrite) the new data link header, update relevant fields in the Layer 3 header, and update other fields in the header (e.g., quality-of-service fields) without the TXP having to reparse the packet header. The transmit process rewrites the data link header, reads the Layer 3 (routing) header from the packet memory and performs the appropriate updates, and then fully constructs the packet by reading the packet payload (body) from packet memory.

The Layer 2 forwarding process in the DECNIS 500/600 operates in a fashion similar to the Layer 3 forwarding process, except that the data link header in the arriving packet is retained from inbound port to outbound port. The Layer 2 forwarding process requires parsing only the data link header from the arriving packet.

5.5   DETAIL SYSTEM ARCHITECTURE

To develop the DECNIS 500/600 as a high-performance switch/router with the technology available at that time, the functional requirements of the switch/router were split into two parts: requirements best handled centrally and those best handled in a distributed fashion.

- **Distributed Functions:** The data link (Layer 2) and network layer (Layer 3) forwarding functions present the main and highest processing load on the actual data path (plane). These functions operate almost totally in a local context, which allows them to be distributed and assigned to a processor associated with a line card or a group of line cards. These data path functions have processing requirements that scale linearly with both network interface speed and number of network interfaces supported by the system. However, some aspects of these per network interface functions, such as initialization of link/line card and processing of control and exception (special) packets, are more efficiently handled only centrally where a more sophisticated processing environment can be created for them. For this reason, these functions can be decoupled from the critical data path processing and placed in a centralized processing module.

- **Centralized Functions:** In contrast to the data plane functions that can be distributed and handled in the line cards, the control and management functions of the system, including the running of the routing protocols and construction of the routing and forwarding tables, are best handled in a centralized module. This is because these are system-wide functions and processes that operate in the context of the switch/router as a whole. The processing requirements associated with these functions are proportional to the size of the network and not to the speed of the network interfaces. Network routing protocols are typically designed to generate a lower amount of traffic (compared to end-user traffic), thereby presenting lower processing loads in routers and switch/routers. The routing protocols have designs that minimize control traffic loads and bandwidth, thereby allowing the construction of relatively simple, low-performance control traffic processing modules in the routing devices.

The above processing considerations resulted in a DECNIS 500/600 architecture that has a central management processor (the MPU) and a set of per line card forwarding processors (for in-place packet forwarding). Each line card communicates on a peer-to-peer basis with other line cards to forward the normal packets transiting the system (that make up the majority of the network traffic).

The management processor behaves, in essence, like a normal full-blown switch/router, except that its participation in packet forwarding is limited to control and management packets. With the system functions properly split between the peer-to-peer line card forwarding processors and the MPU, a buffer and control system was designed to efficiently couple these system processors together.

5.5.1 Centralized System Resources

The DECNIS 500/600 employs three centralized resources: MPC, PRAM, and ARE (Figures 5.3 and 5.4). The decision to centralize these resources was based solely on reducing both the cost and the complexity of the system. The DECNIS

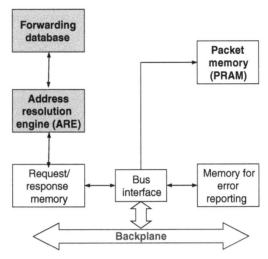

FIGURE 5.4   Packet Memory (PRAM) and Address Resolution Engine (ARE) card.

500/600 was designed to use a partially centralized architecture in which only one processor (the management processor in the MPC) plays the role of the route processor (control engine) as in the traditional router sense. This central processor is responsible for running the routing protocols and building the routing and forwarding tables, network management, and being a central repository for the whole switch/router. The peripheral processors in the line cards will then be responsible for the majority of the actual packet forwarding work.

5.5.1.1 Management Processor Card (MPC)   The MPC is the entity that supports all the routing, control, and management software functionality necessary to tie together the collection of forwarding agents located on the line cards (to form a stand-alone switch/router). The system appears to the rest of the network, from a Layer 3 forwarding perspective, indistinguishable from a traditional router. The processing and memory capabilities of the MPC are designed to be those associated with a typical switch/router.

The MPC (or route processor card) was designed to support two processors: a VAX device that serves as the main processor and a uniprocessor model for the common Futurebus+ based backplane interface. The main MPC processor is responsible for overall control and command of the DECNIS 500/600 system and provides all the control, management, and packet forwarding functions typically found in a monoprocessor switch/router or router. The DECNIS 500/ 600 uses a backplane with 16 bit processor interfaces designed with self-contained functionalities that free the main MPC processor from performing time-critical backplane-associated tasks.

There are a number of (special) packet types (control, management, and exception packets) that are sent directly to the control/management CPU (MPC) for processing. For example, the receive process (RXP) may receive a packet in error or it cannot process a packet with IP header option that requires additional information (i.e., a system context) to process, or a packet that is addressed to the DECNIS 500/600 switch/router itself (including certain multicast packets). The RXP queues such packets to the management processor (control CPU) in exactly the same way it would queue a packet for transmission by a transmit process. The management processor supports a full-function switch/router that is able to handle these special packet types. Similarly, packets sent by the management processor are queued and presented to the appropriate TXP in exactly the same way as an RXP.

When a network interface card receives a packet destined to the MPC, it inspects the packet and informs the MPC whether it is data, routing (Layer 3) control, bridging (Layer 2) control, or system control information (which includes responses from the line card to commands from the MPC). Queues are used at all the network interfaces within the DECNIS 500/600 switch/router. An MPC "assistance processor" identifies the different messages types and queues them on separate internal queues.

Switch/routers must allocate sufficient buffering to handle the routing control messages that they receive and process. Consequently, the DECNIS 500/600 implements a buffer ownership scheme to guarantee the buffering requirements for control messages. To maintain buffering for control messages, the control CPU also implements a mechanism for buffer swapping between the data link and routing modules, as illustrated in Figure 5.5.

The DECNIS 500/600 must guarantee that the data link layer (in the line cards) does not run out of buffers; otherwise, control messages that are regarded as very

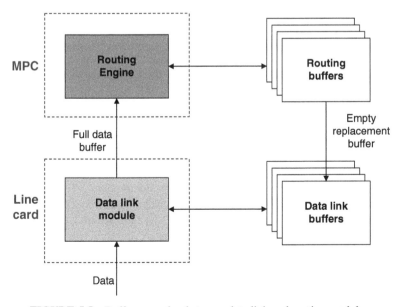

FIGURE 5.5 Buffer swapping between data link and routing modules.

important for routing and control functions will not be received by the MPC. To ensure that the data link layer always has an adequate number of buffers, the control CPU assigns the data link a fixed number of buffers that is maintained at all times. Each time a packet buffer is transferred from the data link module (in the line card) to the routing/control engine (in the MPC), another buffer (a free buffer) is transferred back to the data link module to replace it. If the routing/control engine finds that it has no free buffers to swap, it selects a less important packet, frees up the buffer contents, and passes that buffer to the data link module. This ensures that the data link layer always has buffers for local use. The DECNIS 500/600 line cards also support similar arrangements for line card-to-line card communication.

5.5.1.2   Packet Random-Access Memory   The DECNIS 500/600 was designed to have sufficient shared bus bandwidth to allow the use of a single centralized shared buffer memory (PRAM) for packet storage in the system. This approach, however, causes every incoming packet to cross the shared bus twice, resulting in some loss in bus bandwidth utilization.

5.5.1.3   Address Resolution Engine   The designers of the DECNIS 500/600 analyzed the processing power needed to parse incoming packets and perform the lookup of the destination network address (in a centralized forwarding engine (ARE)) and concluded that the line cards would need some form of assistance if the processing requirements associated with each line card and the ARE was to be made reasonably cost-effective. At the time of the DECNIS 500/600 development, there were already some advancements in the design of hardware forwarding/search engines that made it possible to design a single address parser and lookup engine powerful enough to be shared among all the line cards.

This forwarding engine (ARE) was powerful enough to parse the complex structure of an IP packet (and other protocol packet types) and perform the longest match prefix lookup. In addition, with the DECNIS 500/600 being a multiprotocol switch/router, the forwarding engine was designed to cope with the other routing protocol address formats and the Layer 2 address learning and forwarding requirements of transparent bridging. Also, by centralizing the Layer 2 and 3 forwarding tables, the designers reduced the cost and board area requirements of the line cards and avoided the processing and shared bus (Futurebus+) overhead associated with maintaining several distributed forwarding tables (in the line cards).

Figure 5.6 presents a block diagram of the packet memory (PRAM) and ARE card. As illustrated in this figure, incoming packets and related data and the forwarding tables associated with the ARE are stored in separate dynamic RAM (DRAM) arrays. The ring vector data structures are stored in static memory and are used by the network interface cards to post requests and read responses from the ARE. The ARE was developed as a special ASIC and included some of the other control logic required for the other modules on the card such as the synchronous portion of the Futurebus+ backplane interface and PRAM refresh control.

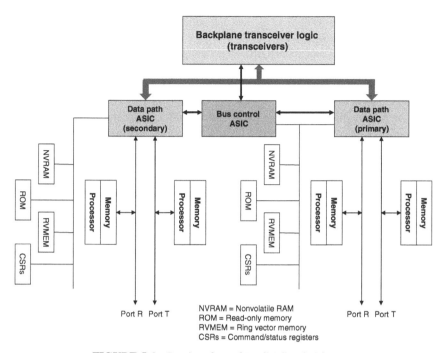

FIGURE 5.6 Bus interface of DECNIS switch/router.

The ARE (which serves as the forwarding engine in the DECNIS 500/600) provides the necessary hardware assistance required in packet parsing and destination address lookups. This single ARE is shared among all line cards in the system. The ARE has sufficient capacity to support a DECNIS 600 switch/router fully populated with line cards that support interfaces each with a bandwidth of up to 2×10 Mb/s. However, beyond this link speed, local destination address (or route) caches have to be used in the line cards. It was determined that the shared bus bandwidth and forwarding table lookup rate required to support multiple fiber distributed data interface (FDDI) line cards would place an excessive processing load on the system. So, to support FDDI line cards, the system was equipped with the central ARE forwarding engine assisted by a line card-resident destination address (route) cache.

5.5.2 Backplane and Interface Logic

The backplanes used in DECNIS 500/600 are based on the Futurebus+ standard (IEEE Standard for Futurebus+: Logical Protocol Specification (IEEE 896.1-1991) and IEEE Standard Backplane Bus Specification for Multiprocessor Architectures: Futurebus+ (IEEE 896.2-1991)). The backplane employs 2.1 V terminated backplane transceiver logic (BTL) (IEEE Standard for Electrical Characteristics of Backplane Transceiver Logic (BTL) Interface Circuits (IEEE 1194.1-1991)). The

DECNIS 500 line cards used 32 bit data and address paths, while the DECNIS 600 backplane was designed to use 64 bits.

In all modules except the PRAM card (which also holds the ARE), the basic backplane interface (shown in Figure 5.6) consists of two ASICs (bus control ASIC and data path ASIC), BTL transceivers, and a number of local memory and registers. These two backplane ASICs are shown in Figure 5.6. The bus control ASIC is responsible for handling bus access requests via a central bus arbitration mechanism, controlling the backplane transceivers and running the parallel protocol state machines for backplane access. The data path ASIC, on the other hand, has two 16 bit processor interfaces (Ports R and T as shown in Figure 5.6), several DMA (direct memory access) channels for each processor port, backplane address decode logic, byte packing/unpacking, and checksum and FCS (frame check sequence) support.

Four DMA channels are provided on the backplane for each processor port. Two of these DMA channels support full-duplex data paths, while the remaining two DMA channels (which are optimized for bulk data transfer) are double-buffered and are configurable to operate in either direction (half-duplex or simplex transmission at a time). As soon as a block fills up, DMA write transfers then occur automatically. Similarly, on suitably configured empty blocks, DMA prefetch reads also occur automatically. The two double-buffered DMA channels (configurable to be half-duplex) are provided in the system to allow bus transactions to take place simultaneously (in parallel) with processor access to the other block.

Data transfers between a processor and any one of the four DMA channels (two full-duplex and two half-duplex channels) are carried out under direct control of the processor, where the processor reads or writes every byte of data to or from the DMA streams. This direct control arrangement provides a great benefit in that it simplifies the design of the related hardware and avoids the need for ASIC DMA support on the processor buses. More importantly, by using processor read and write cycles, the behavior of the system is made deterministic, and furthermore this ensures that the processor has the correct context when all operations have been executed to completion, regardless of the outcome.

The data path ASIC, as illustrated in Figure 5.6, also supports a boot read-only memory (ROM), ring vector memory (RVMEM), command/status registers (CSRs), the geographical address, a local bus containing the control interface for the second bus control ASIC, and nonvolatile RAM (NVRAM) for error reporting. The system is designed to allow a number of the CSRs and the RVMEM to be accessible via the Futurebus+ based backplane.

Another important feature noteworthy is that all resources can be accessed from either of the two 16 bit processor interfaces (Ports R and T). The data path ASIC allows the various subsystems to arbitrate internally for shared resources and is equipped with a number of other features that assist to make data transfers more efficient, for example, a summary register that stores write activity to the RVMEM.

The data path ASIC can be operated/driven from one 16 bit processor interface (Port T). It can be used in the implementation of low-speed line cards that have a relatively simple design. Furthermore, the backplane interface logic contains two (i.e., primary and secondary) data path ASICs (as shown in Figure 5.6) that are served by a common bus control ASIC connected to the primary data path ASIC's local bus. The primary and secondary data path ASICs each takes on a unique device/node identifier in the address space of the backplane. The backplane also provides dedicated lines for temperature sensing, power status, and other system conditions.

5.5.3 Line Cards

The DECNIS 500/600 supports FDDI and Ethernet network adapters as well as synchronous communications interfaces that all use different adaptations of the standard Futurebus+ shared bus backplane interface.

5.5.3.1 Ethernet and FDDI Adapters A number of Ethernet adapters were developed for the DECNIS 50/600. A single-port 10BASE5 (Thickwire) Ethernet adapter was developed that employs a dual-processor architecture (Ports R and T on the primary data path ASIC in Figure 5.6) to interconnect the Ethernet adapter and its associated buffer (tank) memory. This adapter was reworked to place the tank memory interface (TMI) into an ASIC, resulting in a dual-port adapter version. This adapter version was a full remake of the bus backplane interface logic of the DECNIS 500/600 presented in Figure 5.6 but designed to include two Ethernet ports. This adapter was developed in two versions supporting 10BASE5 (Thickwire) and 10BASE2 (Thinwire) Ethernet technologies.

The DEC FDDIcontroller 621 is an FDDI adapter and has an architecture as shown in Figure 5.7. This adapter was developed as a two-module unit to handle the high packet filtering and forwarding rates required for FDDI. The FDDI adapter hardware contains a packet filtering engine (Figure 5.7) closely coupled to a line interface unit supporting FDDI interface functionality, a synchronous interconnect between the two submodules in the cards, and a multichannel DMA engine for data transfer through the adapter.

The DEC FDDIcontroller 621 adapter's DMA engine (in the system interface module (Figure 5.7)) accesses the tank memory under the control of the RISC (reduced instruction set computing) processor, and it can be set up and monitored with very little processor overhead. The system transfers packet data to or from buffers in the packet memory (PRAM) to the tank memory in the line card, where whole packets are maintained in contiguous tank memory address space. A second DMA channel on the network adapter allows the transfers of full packets (in a single burst) to or from the buffer memory located on the line interface submodule of the line card.

The line cards were designed to allow traffic processing between the buffer memory (in the line interface submodule) and the ring vectors to be done in hardware. An additional (third) DMA channel was included to allow prefetching packet header information held in the tank memory (in the system interface

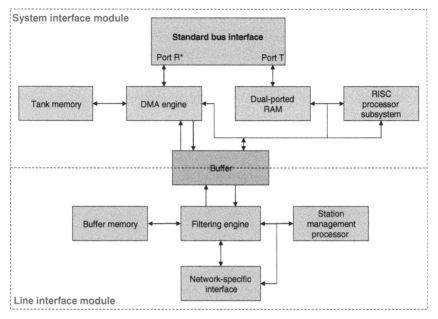

FIGURE 5.7   Line card and system interface architecture.

submodule in Figure 5.7) and then burst transferring it into the RISC processor subsystem for packet processing.

The DMA engine (also shown in the system interface submodule in Figure 5.7) contains a tank memory arbitration mechanism and is capable of queuing multiple commands in addition to operating all DMA channels supported in parallel. The 32 bit RISC processor subsystem (in the system interface submodule in Figure 5.7) is responsible for the line card processing, as well as communicating with the standard bus interface processor via a dual-ported RAM.

5.5.3.2 Synchronous Communications Interfaces   The DECNIS 500/600 supported two synchronous communication wide area network (WAN) adapters. One is a 2.048 Mb/s two-line WAN adapter. The other is a WAN device supporting up to eight lines with each line running a reduced line rate of 128 kb/s. All the lines in two WAN adapters are full duplex with modem control.

The eight-line (128 kb/s) synchronous communication adapter employs a uni-processor architecture plus three industry standard serial communications controllers (SCCs). A 2 m cable connects the clocks and data associated with the lines/channels, as well as one more channel that carries multiplexed modem control information, to a remote distribution panel. Remote distribution panels were designed to support the eight lines (in the eight-line 128 kb/s WAN adapter) using the V.35, EIA422, or RS232 electrical interface standards. A four-line multiple standard panel with a reduced fan-out was also developed that allowed mixed

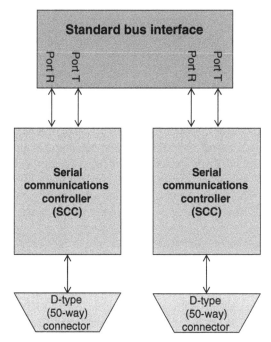

FIGURE 5.8   Block diagram of the DEC WANcontroller 622 adapter.

(V.35, EIA422, or RS232) electrical interfaces from a single synchronous communications WAN adapter. This multistandard remote distribution panel used a 50-pin cable also employed in the other communication products from DEC.

The two-line synchronous communication WAN adapter, on the other hand, employed a four-processor interface, as illustrated in Figures 5.6 and 5.8. The SCC was developed as an ASIC module tailored for the specific form of data-flow processing used in the DECNIS 500/600 architecture. The SCC was closely linked in design to the functioning of the data path ASIC of the backplane interface logic (Figure 5.6) and its associated processors to allow optimal and efficient data transfer.

The design of the two-line WAN adapter hardware resulted in minimal dependency between the data receive and transmit tasks. This was done by recognizing and taking advantage of the limited coupling that exists in acknowledged data link protocols such as HDLC (high-level data link control). The processors exchange state information via a small two-ported RAM located in the WAN adapter's SCC. Two 50-pin V.35 and EIA422 interfaces were provided on the WAN adapter.

5.5.4   Buffer System

Figure 5.9 is a schematic diagram of the buffer system and buffer ownership transfer mechanism of the DECNIS 500/600. The receive processes (RXPs) are allocated only sufficient buffering to handle the data transfer latencies that can occur during the various processing stages of packet forwarding in the system. The longest duration of packet storage (i.e., time being held in buffers) occurs while a packet is owned by the transmit process (TXP).

When a receive process (RXP) completes processing a packet that is destined for a particular destination transmit process (TXP), it swaps the PRAM buffers where the packet is stored for the same number of empty buffers owned by that TXP. It is only when the TXP is able to replace the RXP's packet buffers with empty buffers it owns does the actual buffer ownership transfer take place. If the swap of buffers (between the RXP and TXP) cannot be completed due to the TXP lacking free buffers, the RXP reuses the packet buffers it is currently holding for another incoming packet. This is to prevent a TXP from accumulating buffers and thereby preventing a RXP from receiving and processing incoming packets destined for other output ports.

Designing an efficient buffer ownership transfer scheme was an important part of the development of DECNIS 500/600. The buffer transfer scheme uses a number of single writer/single reader ring vectors (simply referred to as rings), with one ring vector assigned to each RXP and TXP pair (i.e., pairwise) swap of buffer ownership that can occur in the system. With this arrangement, each TXP is assigned one ring vector for each of the RXPs in the system (including the TXP itself), in addition to one ring vector that is reserved for the control CPU (i.e., the management processor).

Whenever an RXP has a packet buffer to exchange (or swap) with a destination TXP, it begins by reading the next (buffer) transfer location in its ring vector (set of pointers) that is associated with (i.e., corresponding to) that TXP. If the RXP finds that the TXP has a free buffer, the RXP swaps that free buffer with the packet buffer it wants to send – thereby keeping the TXP's free buffer as a replacement for the packet buffer. The information involved in the buffer ownership transfer consists of a pointer to the packet buffer, status of buffer ownership, and information that indicates the type of data stored in the packet buffer. The data structure used in the process is a ring vector, and to indicate/specify a ring vector's transfer of ownership, a single-bit semaphore (i.e., to control access to the ring vector) is used.

Figure 5.9 can be used to illustrate how the buffer ownership transfer scheme works. As already described, each transmit process (TXP A or TXP B) has a ring vector dedicated to each of the receive processes in the system (RXP A and RXP B). The process RXP A swaps ring vectors to "Ring A" associated with TXP A and TXP B, and RXP B swaps ring vectors to the "Ring B" associated with TXP A and TXP B.

As the process of buffer ownership transfer takes place, the TXP executes a scavenge process, which scans all the ring vectors associated with it for new packet holding buffers (i.e., buffers with incoming packets). The TXP then queues these packet buffers (specified by the ring vector) in the transmit queues (TXQs). The TXP also replaces the (queued up) entries in the ring vector with entries from its local free buffer list (Figure 5.9). The buffered data-type information enables the TXP in the destination transmit line card to immediately determine the relative priority of the data held in the packet buffer. With this, if the destination line card

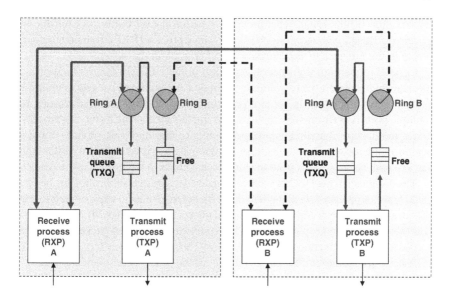

FIGURE 5.9   Buffer transfer scheme: buffer ownership movement.

runs out of buffers due to system overload or congestion, it will be able to discard low-priority packets in order to preserve network stability.

The encodings in the ring vector were optimized to compact the ring vector swap transaction. This was to allow, for all unicast traffic, a single longword read (for new packet carrying buffers) followed by a single longword write (for free buffers) for each buffer ownership exchange or swap. To handle multicast traffic, the DECNIS 500/600 uses a second longword. To reduce the (TXP) processing time associated with running the scavenge process and the amount of traffic that traverses the shared bus of the DECNIS 500/600, the RAM in which the ring vectors are held is placed on the transmit side of a line card. Appropriate hardware is included to watch for activity in the ring vectors so that they can be reported to the TXP.

To allow efficient use of PRAM resources, long packets are fragmented and stored over a number of buffers. The unit buffer size (of 256 bytes) achieved a better compromise between memory use efficiency and processing overhead associated with buffer management. This buffer size compromise may lead to a fraction of the arriving packets stored over more than a single buffer. This means that when an output port on a line card goes into congestion, it is no longer certain that a complete set of packet buffers (for a full packet) will be swapped.

Therefore, the system ensures that a processed packet is queued for transmission in the transmit queues (TXQs) only if it has been fully transferred to the transmit process (TXP) on the destination line card. To handle dissimilar scavenge and swap process speeds, the transfer of buffers (between an RXP and a TXP) is staged using a process called binning. In this process, a TXP obtains a complete set of packet buffers (holding a full packet) from an RXP before queuing the packet for transmission in the TXQs. This prevents a partial packet buffer transfer due to system overload (or congestion) or a slow receive process (RXP) from impeding or blocking the progress of data transfer on other ports in the system.

Layer 2 forwarding (transparent bridging) in the DECNIS 500/600 requires a mechanism to support the forwarding of flooded and multicast packets to multiple destination output ports. In some architectures, the flooding and multicasting to multiple ports is performed by replicating a single packet using a copying process. Other architectures handle this process using a central multicast service. Using a central multicast service can result in synchronization issues when a destination Layer 2 address transitions from the unknown address to the learned address state.

The designers of the DECNIS 500/600 realized that packet replication by the line cards was not practical since the line cards do not keep a local copy of an arriving packet after it has been copied to the central packet memory (PRAM). The solution was therefore to employ a system in which multicast designated packet buffers are loaned to all the destination transmit line cards. The system used a "scoreboard" (that indicates outstanding buffers loaned) to record the state of each multicast buffer.

When a loaned multicast buffer is returned from all its destination transmit line cards, the buffer is appended/added to the multicast free buffer list (or queue) and becomes available for reuse. The multicast buffer loan and return processes are similar to the normal unicast packet buffer scavenge and swap process described above, with the addition that the ring vector used is extended slightly to include the information needed for rapidly getting the data that are stored in the memory location pointed to by the ring vector (pointer).

5.5.5   DECNIS 500/600 Software Architecture

In this section, we discuss the software architecture of the DECNIS 500/600. We discuss first the software architecture of the management processor (in the MPC) followed by the software architecture of the receiver and transmitter in a line card. The discussion is followed by details on how the system forwards multicast packets.

Figure 5.10 illustrates the software architecture of the MPC in the DECNIS 500/600. The software allows the MPC to act as a full-function switch/router and X.25 gateway on its own. The architecture also includes adaptation software layer necessary to hide the details of the DECNIS 500/600 MPC software environment from the details of the line cards.

The control and management part of the system software code includes the Layer 3 (routing), Layer 2 (bridging), network management, and X.25 software. This code was derived from the software used in the Digital's WANrouter 500 (an extended version of the software used in that device). The software extensions included in the DECNIS 500/600 code version were important to supply Layers 2 and 3 forwarding table updates and system configuration information to the adjacent DECNIS 500/600 environment adaptation module. The environment adaptation module hides the control and system management module from the details of the forwarding functionality in the line cards.

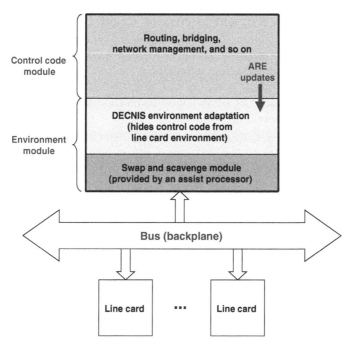

FIGURE 5.10   Software structure in the management processor card.

The DECNIS 500/600 environment adaptation module (Figure 5.10) contains key system software code components that are responsible for line card data link initialization, line card control code, and the software code needed to transform the Layer 2 and 3 forwarding table updates into appropriate data structures used by the ARE. The DECNIS 500/600 environment adaptation module has another important software component that handles the scavenge and swap packet buffer functions required for communication with the RXPs and TXPs in the line cards. These scavenge and swap functions have real-time operating constraints associated with them, thereby requiring them to be split between an assist processor (not depicted) and the management processor on the MPC.

The control and management software code module (Figure 5.10) was designed to allow the management processor in the MPC to function as full-fledged switch/router and also to allow the introduction of new functionality in the DECNIS 500/600 in stages. Whenever the need for a new protocol type in the system arises, this feature can be initially implemented and executed in the management processor, and with the line cards providing the required data link framing and network interface service. This allows the DECNIS 500/600 at a later point to move the packet forwarding functions (initially implemented in the MPC) to the line cards in order to provide a more enhanced system packet forwarding performance.

5.6   UNICAST PACKET RECEPTION IN A LINE CARD

Figure 5.11 shows the processes that run on the receive side of a network interface card. The four main processes are as follows:

- Receive process (RXP), which is the main process and aspects of which have already been described.
- Receive buffer system ARE process (RXBA), which receives inputs from the preaddress queue.
- Receive buffer system descriptor process (RXBD), which receives inputs from the receive bin queue.
- Swap process, aspects of which have also already been described.

The RXP (receive process) continuously polls/interrogates the link communications controller in the line interface until it notices an arriving packet.

The RXP then fetches a pointer to a free buffer in the packet memory (PRAM) from the free buffer list queue. The RXP parses the data link header and the IP header from the arriving packet, and copies the packet byte-by-byte (as the header information is parsed) into the free buffer. The RXP determines from the data link header if the arriving packet should be Layer 3 forwarded (routed) or Layer 2 forwarded (bridged).

Once the RXP has decided the type of forwarding required for the incoming packet, the IP destination address or the destination MAC address is copied to the ARE, in addition to some information to indicate to the ARE which forwarding table to use for the lookup. The ARE incorporates some hardware to assist in the Layer 2 address (transparent) bridge learning process.

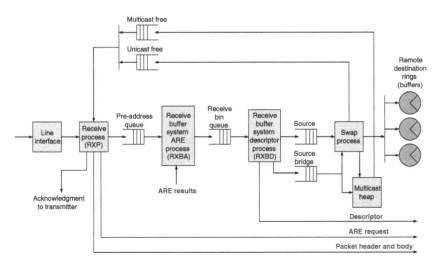

FIGURE 5.11   Network interface card receive processing.

To prevent the ARE assisting hardware from accidentally learning an incorrect MAC address, the ARE is only allowed to start a MAC address lookup (in its Layer 2 forwarding table) after the RXP has completely received the packet and established that it carries a correct checksum. This restriction does not apply to Layer 3 addresses, which the ARE can look up in its Layer 3 forwarding table even before a full packet has been received, thus reducing data transfer latency.

If the packet is to be Layer 3 forwarded (the data link header is discarded), only the IP header and the packet body are copied to the buffer(s) in packet memory. The system also stores for later use the source MAC address of the incoming packet or, when the packet arrives on a multichannel line card, the channel (identifier) on which the packet was received. The system stores as well several other protocol-specific items for the packet. All this packet information is later used by the system to construct the packet descriptor.

The RXP stores the buffer pointer in the preaddress queue (shown in Figure 5.11) until the pointer can be reconciled with the result of the packet's Layer 3 destination address lookup (in the ARE's forwarding table). In the case where line card processing involves data link protocols such as HDLC (that use acknowledged data transfers), the RXP exports the latest acknowledgment status to the destination TXP.

The ARE is also polled continuously by the RXBA (receive buffer system ARE process) for the result of the forwarding table lookup (for a particular destination address), after which the RXBA stores the result in an internal data structure associated with the corresponding packet. The RXBA then moves the buffer pointer along with any additional buffer pointers to other packet buffers (used to store the remaining parts of a bigger packet) to the receive bin queue (shown in Figure 5.11).

The RXP, RXBA, and ARE forwarding engine destination address lookup processes and the link interface transmission process all operate asynchronously. This thus creates a system design in which a number of ARE forwarding table lookup results can be pending and that can also have indeterministic completion time.

The above issues can cause the reconciliation of forwarding table (destination address) lookup results for arriving packets and their associated buffers in packet memory to take place before or even after a whole packet has been received. Given that an error could occur in the arriving packet, the system was therefore designed to not take any further action on the packet until the whole packet has actually been received and all its packet buffers have been moved to the receive bin queue as shown in Figure 5.11.

The above staging process was used to avoid providing in the system a complex abort mechanism to clear/remove erroneous packets from the scavenge, swap, and transmit processes. The system was designed to have the RXBA poll the ARE at a rate of exactly one poll per forwarding table lookup request when operating under normal traffic load. With this arrangement, an ARE poll failure can increase the backlog in the preaddress queue (which holds the packet buffer pointers), but does not cause it to grow beyond two packets.

This mechanism also minimizes the shared bus (Futurebus+) bandwidth wasted in unsuccessful (or failed) ARE poll operations. However, when the line card receiver (RXP) becomes idle (with no new packets arriving), the ARE poll rate increases and the outstanding packets (in the preaddress queue shown in Figure 5.11) are rapidly processed to clear the queued/backlogged packets.

The RXBD (receive buffer system descriptor process) records the packet descriptor into the foremost part/front of the first packet memory (PRAM) buffer that is used to store the packet. The packet descriptors are protocol-specific (containing all relevant protocol information required to forward the packet) and require a callback into the corresponding protocol code to construct the descriptor.

After the packet descriptor is written into place, the buffer pointers (associated with the processed packet) are transferred to the source queue (shown in Figure 5.11), ready for forwarding (by the swap process) to the destination network interface card. The packet buffer is then swapped with the TXP in destination network interface card and the free buffer resulting is added to the free buffer queue.

5.7 UNICAST PACKET TRANSMISSION IN A LINE CARD

The transmitter functions of a network interface card are shown in Figure 5.12 and consist of five processes:

- Scavenge rings process, which takes inputs from the destination rings.
- Scavenge bins process, which takes inputs from the destination bins.
- Transmit buffer system select process (TXBS), which takes inputs from the holding queues.

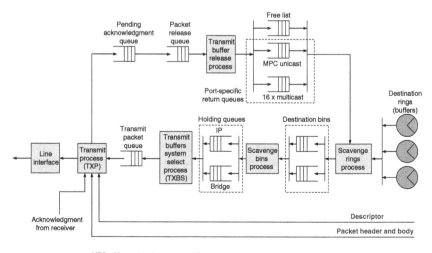

MPC = Management processor card

FIGURE 5.12   Network interface card transmit processing.

- Transmit process (TXP), aspects of which have already been described.
- Transmit buffer (TXB) release process, which takes inputs from the packet release queue.

The scavenge rings process is the transmit-side entity that examines/scans the swap rings (i.e., ring vectors holding pointers) for new packet buffers waiting to be queued and transmitted out of the switch/router. The scavenge process then replaces these new packet buffers with free buffers. The new packet buffers are then stored/ queued in destination bins that serve as reassembly bins (one destination bin for each destination ring) in order to facilitate only full/complete packets to be queued in the subsequent holding queues.

The scavenge process first attempts to refill the destination rings with free buffers from the port-specific return queues, and if this fails, it takes the free buffers from the free list. The port-specific return queues (i.e., the MPC unicast and 16× multicast queues in Figure 5.12) are used primarily for multicast packet forwarding. The scavenge bins process inspects the destination bins to determine if they hold full packets and moves them to the appropriate (IP, bridge) holding queues. The packets are queued in the holding queues by protocol type, since different protocols have different packet fields, protocol processing requirements, and traffic characteristics.

Using round-robin scheduling, the TXBS process schedules the packets from their holding queues. The round-robin scheduling is to prevent protocols (e.g., TCP) with dynamic congestion control algorithms from being driven into congestion timeout by other protocol packet types with no congestion control mechanisms. This also allows both Layer 2 and Layer 3 protocol traffic to have a fair share of the link resources when traffic overload occurs.

The scavenge bins and TXBS processes together implement any necessary congestion-related dropping and marking of congestion bits in packets during congestion in addition to any packet aging functions. By allowing the queuing time of packets to be minimal in the receiver (side of a line card), the DECNIS 500/600 designers were able to simplify the congestion control algorithms and allow them to be executed in the transmit path.

The TXBS process selects a packet and transfers it to the TXP (via the transmit packet queue), which then transmits it to the network. TXP writes the appropriate information from the packet descriptor, rewrites the IP header, and prepends the data link header. When the TXP transmits a packet belonging to a protocol like HDLC (that uses explicit data acknowledgments), the packet is transmitted but also transferred to the pending acknowledgment queue where it waits for an acknowledgment message from the remote end.

Before the TXP transmits each packet requiring explicit acknowledgment, it examines the current acknowledgment state for the packet indicated by the remote end (receiver). The TXP may transfer packets requiring acknowledgments from the pending acknowledged queue to the packet release queue, and if it receives a retransmission request for the packet, move the packet back to the transmit packet queue.

The TXB release process removes packets (buffers) from the packet release queue and splits them into a series of queues to be used by the swap process. Buffers associated with unicast packets are placed in the free buffer list (i.e., free pool in Figure 5.12). Buffers associated with multicast packets are returned to the port-specific queue for the source network interface card, and ready to be assigned to their originating receiver (TXP). Buffers associated with packets targeted for the control and management CPU (MPC) are also queued separately in their port-specific queues.

5.8 MULTICAST PACKET TRANSMISSION IN A LINE CARD

A packet buffer associated with Layer 3 multicasting and Layer 2 multicasting or flooding must be transmitted by a number of line cards. This process requires the system to swap a special type of ring vector, which indicates that the buffer (for the multicast data) is only on loan to the transmitting line card, and upon completion of packet transmission must be returned to its receiver (i.e., buffer owner).

In addition to the information carried in a normal ring vector (i.e., normal packet type, buffer identification information, and fragmentation), the special (or multicast-associated) ring vector carries local referencing information, which indicates where it (the ring vector) is stored on the multicast heap. The receiver (owner) maintains a list/record of which multicast packet buffers are on loan to which transmitting line card. The scavenge process must also keep a record of which ring vector it found a particular pointer.

After transmitting a multicast packet, the TXB release process returns the ring vector to the associated (or owner) port-specific return queue. The returned ring vectors are then eventually transferred to their receiving line card (owner) through the swap process. As the owner receives these returned multicast buffers, it marks them off against the record of multicast buffers it loaned out. As soon as the receiver gets back a buffer from all transmitting line cards to which it was loaned, the buffer is returned to the free list. To allow this process to work successfully, the system reserves some buffers specifically for multicast packet forwarding.

6

ARCHITECTURES WITH BUS-BASED SWITCH FABRICS: CASE STUDY— FORE SYSTEMS POWERHUB MULTILAYER SWITCHES

6.1 INTRODUCTION

This chapter describes the architectures of the Fore Systems PowerHub multilayer switches [FORESYSWP96]. The PowerHub multilayer switches (PowerHub 7000 and 6000) perform both Layer 2 and Layer 3 forwarding and support a wide range of physical media types for network connectivity. The PowerHub supports Layer 3-based VLANs, IP routing (RIP and OSPF), and IP multicast among other features.

The PowerHub employs a software-based packet processing engine (with multiple RISC processors) for both route processing and packet forwarding. The main argument for this software-based solution at the time the architecture was proposed is that it is very flexible to allow routing and management protocols, packet forwarding features to be added or updated, and software bugs to be fixed with simple software fixes, upgrades, or download.

It was viewed at that time that network devices (switches, routers, switch/routers, etc.) that use ASICs to process and forward packets are fixed-function devices that do not have the kind of flexibility required when the system requires enhancements or modifications. As an example, when the designer requires the switch to be enhanced with IEEE 802.1Q VLAN tagging for Ethernet frames, most ASIC-based switches will require parts of the switch to be replaced with new ASICs incorporating this new tagging feature. However, the PowerHub, being software-based, allows the VLAN tagging feature to be added to its software base with no hardware swaps required.

Switch/Router Architectures: Shared-Bus and Shared-Memory Based Systems, First Edition. James Aweya.
© 2018 The Institute of Electrical and Electronics Engineers, Inc. Published 2018 by John Wiley & Sons, Inc.

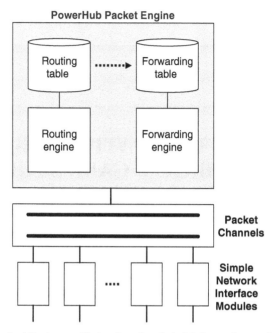

FIGURE 6.1   Architectures with bus-based switch fabrics and centralized forwarding engines.

As observed in [FORESYSWP96], the PowerHub routing and forwarding software was designed as a fully optimized software platform to take advantage of the semiconductor device and hardware speed improvements available at that time. A design goal was to allow this innovative software architecture to effectively use capabilities such as shared memory, cache organization, write buffers, and burst-mode CPU transactions.

Although some of the components and features of the PowerHub are obsolete and will not be used in today's architectures and networking environments, they are discussed in this book to allow the reader to appreciate how multilayer switches have evolved over the years.

Adopting the architecture categories broadly used to classify the various designs in Chapter 3, the following architectures are covered in this chapter:

- Architectures with bus-based switch fabrics and centralized forwarding engines (see Figure 6.1):
  - Fore Systems PowerHub multilayer switches with Simple Network Interface Modules (SNIMs)
- Architectures with bus-based switch fabrics and distributed forwarding engines (see Figure 6.2):
  - Fore Systems PowerHub multilayer switches with Intelligent Network Interface Modules (INIMs)

PowerHub Packet Engine

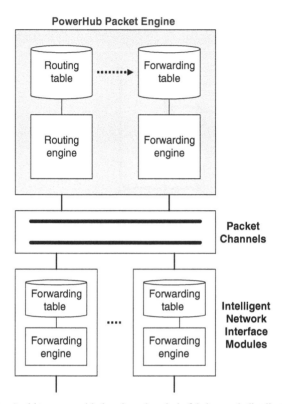

FIGURE 6.2   Architectures with bus-based switch fabrics and distributed forwarding engines.

6.2   POWERHUB 7000 AND 6000 ARCHITECTURES

The PowerHub has an architecture that supports a shared memory (i.e., the packet buffer memory), program memory, flash memory, multiple RISC processors, network interface controllers, packet switching engine, and an ASIC for optimizing shared memory access. All packets in transit through the switch are first received into the shared memory, processed by the CPUs (the RISC processors) and then forwarded to the appropriate destination port(s) for transmission to the network. The shared memory architecture was designed with the goal of simplifying the packet processing and forwarding algorithm. Figures 6.3–6.7 show the main architectural features of the PowerHub switches.

To improve the packet forwarding performance and scalability of networks that deploy the PowerHubs, the multiple RISC processors are distributed throughout the system and, in addition, the switch uses a combination of centralized and distributed processing. The Fast (100 Mb/s) Ethernet interface modules, Fiber Distributed Data Interface (FDDI) modules, and Asynchronous Transfer Mode (ATM) modules all

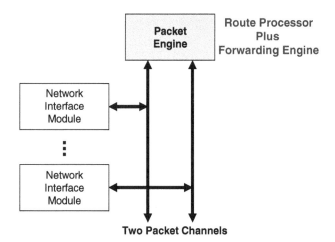

FIGURE 6.3 PowerHub 7000 system architecture block diagram.

implement locally their own Layer 2 and Layer 3 forwarding functions, thus increasing both the packet forwarding performance and scalability of the switch.

Packet forwarding decisions in the PowerHub can be performed at the Layer 3 (IP) or at Layer 2. The PowerHub can also process and forward packets based on user-defined filters at either Layer 2 or Layer 3. In addition, the PowerHub supports network management features such as statistics gathering, security filtering, and port monitoring (also known as port mirroring).

Port monitoring can be used to capture/copy network traffic passing through a PowerHub port for analysis. This feature allows packets passing through a port on the switch to be copied to another port on the switch that has a connection to a traffic probe device, a Remote Monitoring (RMON) probe, or a security device. Essentially,

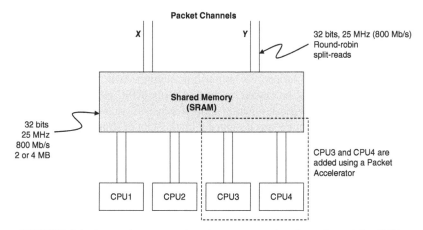

FIGURE 6.4 Interaction among shared memory, packet channels, and the CPUs.

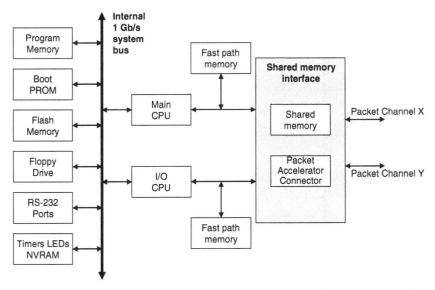

FIGURE 6.5 Block diagram of the PowerHub 7000 system architecture with multiple processors.

port monitoring allows packets in the transmit, receive, or both directions on one source port (or more) to be copied or mirrored to a destination port for analysis.

6.2.1 PowerHub 7000 Architecture

In the PowerHub 7000, the Network Interface Modules (NIM) are connected to the packet engine through a set of buses as shown in Figure 6.3. The PowerHub 7000 architecture also supports the following:

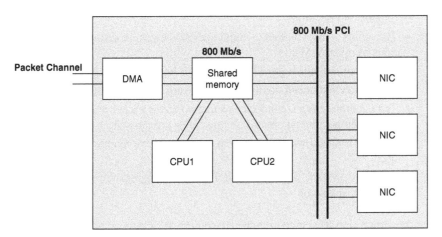

FIGURE 6.6 Block diagram of an Intelligent Network Interface Module.

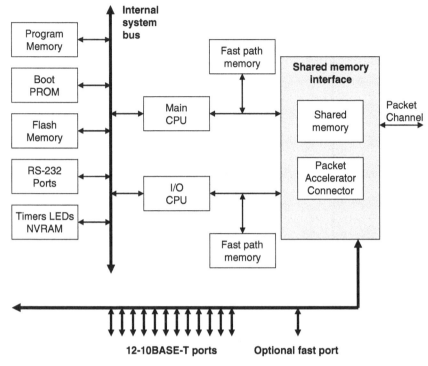

FIGURE 6.7 Block diagram of the PowerHub 6000 system architecture.

- Two 800 Mb/s packet channels (resulting in a total of 1.6 Gb/s bus bandwidth) as shown in Figure 6.3.
- A 10-slot chassis to hold system modules.
- Hot-swapping of NIMs (i.e., the replacement of a NIM while the system remains in operation).
- The ability to reset/initialize individual NIMs without disrupting the operations of other NIMs.
- Up to four redundant power supplies with load-sharing capabilities.
- Up to three redundant cooling fans for every five NIM slots in the chassis.
- Temperature sensors (connected to alarms) on the packet engine to sense and trigger temperature alarms when thresholds are crossed.
- Automatic shutdown of the system when excessive temperature conditions occur.
- Auxiliary system buses for FDDI network aggregation or concentrator modules, and so on.

The PowerHub 7000 supports a shared memory, RISC processor-controlled route processing, and packet forwarding component. All packets arriving to the system are stored in a centralized shared memory. The relevant packet fields are inspected and processed by RISC processors to make the forwarding decisions required. After determining the forwarding information, the packets are then sent to the appropriate destination port or ports (for multicast traffic) for transmission to the network.

To achieve scalable performance, the PowerHub distributes the processors and shared memory to other system modules that include the INIMs, which consist of 100 Mb/s Ethernet, FDDI, and ATM modules. Distributing the processing and shared memory to the INIMs allows processing and forwarding of packets locally within the INIMs.

The main components of this multilayer switch are described below (see Figures 6.4–6.6). At the time the PowerHub was designed, Fast (i.e., 100 Mb/s) Ethernet and FDDI were considered high-speed networking technologies with FDDI mostly used as the preferred transport technology for network backbones.

6.2.1.1   The Packet Engine: Combined Route Processor and Forwarding Engine   The packet engine module holds the "intelligence" of the PowerHub and consists of a number of processors and a shared memory, other components that perform high-speed Layer 2 and Layer 3 forwarding, and network management functions.

- **Shared Memory:** The packet engine supports 2 MB of multiport shared memory, which provides 800 Mb/s of bandwidth (i.e., 32 bits wide memory clocked at 25 MHz). The shared memory supports four ports with one port assigned to each of the two packet channels and one port to each of the two processors. When a packet accelerator is installed in the system, an additional 2 MB of shared memory is added, in addition to two CPUs (see Figure 6.4).

    All packets arriving to the PowerHub are received by all the NIMs and stored into the shared memory except for the INIMs (i.e., ATM, FDDI, and 6×1 Fast Ethernet modules). The received packets are inspected by the processors and marked accordingly for Layer 2 or 3 processing and then forwarding as required. Processed Layer 2 or 3 packets are read directly by the NIMs from the shared memory and transmitted to the destination ports as specified by their forwarding instructions.

    The shared memory is designed from standard cached SRAM chipsets. Custom-designed bus/memory interface ASICs are used to provide the multiple ports required for the processors and packet channels to access the shared memory. While other LAN switches at the time the PowerHub was designed may use a shared bus arbitration approach, the PowerHub employs a pipelined, shared memory access approach. This pipelined approach to the shared memory allows each shared memory port to operate as if it has exclusive access to the shared memory.

- **Multiple Processors:** As illustrated in Figure 6.4, the PowerHub architecture supports multiple CPUs that communicate via the shared memory. Two CPUs are used in the PowerHub 7000 packet engine, and two additional CPUs (shown in Figures 6.4 and 6.5) are on the packet accelerator (when installed on the system).

- **Main CPU:** The main CPU (see Figure 6.5) runs the Layer 2 and 3 packet forwarding algorithms in the system. This CPU also runs the routing protocols to maintain the main routing table and forwarding table. It also handles the network management functions in the system. If the optional packet accelerator (with two extra CPUs) is installed in the system as illustrated in Figure 6.5, the packet forwarding code is run on one of its CPUs and the management and overhead tasks on the other CPU.

- **Input/Output (I/O) Processor:** The I/O CPU (see Figure 6.5) handles real-time processing functions for the NIMs, some of which include NIM initialization, buffer management, packet reception, packet transmission, and error handling. When the optional packet accelerator is used, the I/O Processor functions are split and handled by two CPUs. Packet reception and buffer cleanup are handled by one CPU, while packet transmission is handled by the other. Each of these CPUs has access to 128 kB of private (or "fast path") memory that holds the performance-critical software code and related data. Each CPU also supports an additional 20 kB of on-chip cache memory.

- **Memory Resources:** The PowerHub packet engine supports other memory types used by both CPUs (see Figure 6.5) in addition to the shared memory used for storing packets:
  - 24–40 MB DRAM as main memory for PowerHub system software.
  - 512 kB EEPROM as boot memory.
  - 8 kB battery backed-up RAM as nonvolatile memory.
  - 4 MB flash memory for software code storage.

- **RS-232 Ports:** The packet engine supports two RS-232 ports each with these capabilities:
  - Asynchronous modem control lines supported: TXD, RXD, DTR, DCD, CTS, and RTS.
  - The modem speeds supported in bits/s are as follows: 1200, 2400, 4800, 9600, and 19,200.

  The RS-232 standard was renamed as EIA RS-232, then EIA 232, and is now referred to as TIA 232. The RS-232 ports also allow a network administrator to directly access all the PowerHub system functions through a command-line interface (CLI). These ports also allow the network administrator to perform in-band network management via Telnet and SNMP.

- **ID PROM and Temperature Sensor:** To allow for self-inventory and technical support, the PowerHub leaves the factory with part of its boot PROM programmed with the packet engine's serial number, part number, and revision number, in addition to other useful device information.

  In addition, the packet engine supports alarms and temperature sensors that can be read by the system management software to determine if the temperature of the packet engine is within acceptable limits. These readings from the temperature sensors can trigger alarms that can be at the board or system level.

- **Expansion Connector:** The packet engine supports an expansion connector for adding the optional packet accelerator to the PowerHub (Figures 6.4–6.7). The packet accelerator is a performance-enhancing daughter module that contains two additional CPUs and 2 MB of shared memory.

- **Packet Accelerator:** The packet accelerator does not use a NIM slot; instead, it connects directly to the packet engine through the expansion connector. The packet accelerator supports two CPUs that are identical to the CPUs used by the packet engine. Each of these CPUs is given a 128 kB of private "fast path" memory and 20 kB of cache memory. In addition, the packet accelerator is also assigned 2 MB of shared memory, resulting in a total of 4 MB of shared memory.

  When a packet accelerator is used in PowerHub 7000, its packet processing and forwarding capacity increases by approximately 50%. By using the packet accelerator, both the processing power and the amount of shared memory are increased.

- **Packet Channels:** Each of the two packet channels (Figures 6.4 and 6.5) can support a peak bandwidth of 800 Mb/s (i.e., 32 bits wide channel each clocked at 25 MHz). The efficiency of the packet channel depends on the type of NIM using that packet channel. Reference [FORESYSWP96] states the efficiency of the packet channel to be 55–75%, which makes the effective capacity of a packet channel to be 440–600 Mb/s.

6.2.1.2 Network Interface Modules   The PowerHub 7000 supports several types of NIMs for network connectivity:

- **Simple Network Interface Modules (without Forwarding Engines):** In the simplest configuration, the PowerHub 7000 can support Simple NIMs (SNIMs) that do not have local packet processing and forwarding capabilities. These SNIMs forward all packets received from the network (and in transit through the PowerHub) to the packet engine for all processing and forwarding decisions. Packets in transit received by a SNIM are transferred over the packet channel by SNIM chips and stored in the shared memory located on the packet engine. The header information of the stored packets is then examined by the packet engine's CPUs and forwarded, as required. Furthermore, packets that have completed the forwarding table lookup processed are read from the packet engine's shared memory, transferred over the packet channel to the appropriate outbound NIM and transmitted to the network by the NIM chips.

- **Intelligent Network Interface Modules (with Forwarding Engines):** The PowerHub 7000 also supports Intelligent NIMs (INIMs) that have their own shared memory, CPUs, Layer 2 and 3 forwarding functions, and system management software (Figure 6.6). These INIMs implement a distributed packet processing, forwarding, and shared memory architecture, which allow the PowerHub to scale in packet forwarding capacity and performance. The FDDI, Fast Ethernet, and ATM modules in the PowerHub 7000 are all designed as INIMs. These network technologies were all considered high-speed technologies at the time the PowerHub was designed.

The INIMs and the packet engine are similar in architecture and share similar features. INIMs have two RISC CPUs, a shared memory for storing, locally, packets, and a local program and data memory. The local forwarding tables used by the INIMs for forwarding decisions are generated by the packet engine that runs the routing protocols.

This distributed packet forwarding and shared memory architecture implements a centralized route processing (or control) engine in the packet engine coupled with distributed packet forwarding in the INIMs. In this architecture, the INIM has a copy of the main Layer 2 and 3 forwarding tables that are maintained by the packet engine. All updates to the packet engine's Layer 2 and 3 forwarding tables are also transferred to the INIMs local Layer 2 and 3 forwarding tables.

Packets received by an INIM are stored locally in its shared memory and a forwarding decision is performed locally by the INIM's CPU using its local copy of the forwarding tables. This allows the PowerHub to distribute the forwarding intelligence required for packet forwarding to the INIMs, thereby significantly increasing the packet forwarding capacity of the whole system. The local forwarding tables in the INIMs are always kept synchronized to the main forwarding tables in the packet engine.

- **PowerCell ATM Modules:** The PowerCell ATM networking modules of the PowerHub are considered part of the INIM family. The PowerCell supports, locally, its own Segmentation and Reassembly (SAR) chip, shared memory, two CPUs, and a local program and data memory. The PowerCell can be used in multiple networking scenarios. For example, it can be used as a "one-armed" router to forward inter-VLAN traffic. Note that traffic from one Layer 2 VLAN to another has to go through a router. The PowerCell can also be used as a multilayer edge switch or Layer 2 and 3 connected to an ATM backbone.

- **FDDI Concentrator Modules:** The PowerHub FDDI concentrator allows multiple stations to share the bandwidth of a 100 Mb/s FDDI ring. A PowerHub FDDI switch module (not discussed here) acts as a switch (connected to the packet engine through the packet channel) to which up to four FDDI concentrator modules can be connected. The FDDI concentrator modules connect to the FDDI switch module through one or two FDDI rings. The FDDI switch module in turn connects to the packet engine via the packet channel.

6.2.2 PowerHub 6000 Architecture

Compared to the PowerHub 7000, the PowerHub 6000 is a smaller, compact design with a packet engine and two optional NIMs in a three-slot chassis (Figure 6.7). This design supports 12 10BASE-T Ethernet ports and optional ports for Fast Ethernet, FDDI, and ATM connectivity. PowerHub 6000 also supports a main power supply plus an optional, secondary, or redundant power supply. It also supports the full range of routing, management, and system software features found on the PowerHub 7000.

The PowerHub 7000 and the PowerHub 6000 are architecturally similar but the latter differs by supporting the following features:

- One packet channel to which the NIMs are connected.
- 12 10BASE-T Ethernet ports, and an optional FDDI port with connectivity to the packet engine, through its own internal packet channel.
- Does not support hot-swapping of modules.
- A 600 Mb/s shared memory bandwidth that is slightly lower than the shared memory in the PowerHub 7000.
- Physically smaller NIMs.
- Maximum configuration supports a smaller number of modules.

In the PowerHub 6000, unlike in the PowerHub 7000, packet filtering and forwarding for traffic arriving at FDDI NIMs are performed by the packet engine's CPUs in the PowerHub 6000's, and not by local CPUs on the INIMs (as in PowerHub 7000). The operation of PowerHub 6000, other than those cited above, are almost identical to that of PowerHub 7000 [FORESYSWP96].

6.3 POWERHUB SOFTWARE ARCHITECTURE

The PowerHub supports four CPUs (two on the packet engine and two on the packet accelerator (Figure 6.4)) with its software designed to run in a multiprocessor system. The design goal is to allow the system to run with as little overhead as possible. The sources of overhead in a multiprocessor system, traditionally, are context switching, interrupt processing, and locking/unlocking data structures.

Traditionally, interrupts and context switching following interrupts are often employed in software-based switching and routing systems developed to handle unpredictable network traffic loads. For instance, when a packet arrives at a network interface, a hardware interrupt is asserted to alert the appropriate software processing module to handle the packet. In the meantime, the system does not use any processing time to look for nonevents in the systems. Such designs are most desirable or suitable when the packet arrival rate to the system is low.

An alternative to interrupt-driven software designs is the polling software architecture used in the PowerHub [FORESYSWP96]. Here, when the PowerHub has detected that a port is active (i.e., "Link detected" on a port), that port is added as part of the packet polling loop maintained by the system. When a port is polled, its packets are transferred over one of the two 800 Mb/s packet channels and stored in the packet engine's shared memory for packet processing.

The packet processing tasks (for Ethernet frame forwarding example), in general, include the following for both centralized packet forwarding in the packet engine and distributed packet forwarding in the INIMs [FORESYSWP96]:

1. Check if a packet has arrived.
2. Check if the packet has errors (e.g., Frame Check Sequence (FCS) errors, size errors, packets tagged for a specific VLAN on a port not configured for that VLAN, etc.).
3. Update the PowerHub packet receive statistics.
4. Check if the packet is to be Layer 2 forwarded (bridging) or Layer 3 forwarded (routing).
5. If packet is to be Layer 2 forwarded, then extract the destination MAC address and perform the transparent bridging algorithm:
   - **Forwarding:** If the destination MAC address is in the Layer 2 forwarding table, then forward the packet out the port and VLAN that is associated with that destination MAC address.
   - **Filtering:** If the port to forward the packet is the same port on which it arrived, then discard the packet (no need to waste resources forwarding the packet back out its originating port).
   - **Flooding:**
     - If the destination MAC address is not in the Layer 2 forwarding table (implying the address is unknown), then forward the packet out all other ports that are in the same VLAN as the packet (no need to waste resources flooding the packet out the same port).
     - If the destination MAC address of the packet is FFFF.FFFF.FFFF (the broadcast address), then forward the packet out all the other ports that are in the same VLAN as the packet.
6. If packet is to be Layer 3 forwarded, extract the IP header information and perform the following:
   - Perform Layer 3 forwarding table lookup to determine next hop IP address and its associated (egress) port (SNIM or INIM).
   - Update time-to-live (TTL) field.
   - Recompute IP header checksum.
   - Update source and destination MAC addresses in the outgoing Ethernet frame.
   - Recompute Ethernet frame FCS.
7. Check for exception and special case packets: ARP, Spanning Tree Protocol (STP), ICMP, IGMP, RIP, OSPF, Telnet, SNMP, and so on.

8. Forward the processed packet to the appropriate SNIM or INIM.
9. Allocate fresh buffers from shared memory to the receive SNIM or INIM.
10. Update the PowerHub packet forwarding statistics.
11. Check if transmission is completed.
12. Check for transmission errors.
13. Update PowerHub transmission error statistics.
14. Flush or reclaim the buffers used for the recently transmitted packet.

6.3.1 PowerHub Polling Architecture

A key component on which the PowerHub polling architecture depends to carry out its functions is the PowerHub network interface (or NIM) controllers. In the traditional design, a CPU steps in immediately as soon as the interface controller indicates that a packet has been received or is to be transmitted. In the polling architecture, the interface controller does not require the CPU to respond immediately when a packet has arrived or requires transmission.

The PowerHub polling uses a memory-based descriptor structure that allows the interface controller to store the status of a received packet or a packet requiring transmission. This allows the interface controller to move on to the next packet without CPU intervention. This architecture allows up to 512 packets to be received or queued for transmission back-to-back.

This means that when many packets are queued for transmission due to heavy load on a network from a PowerHub switch 10BASE-T port, for example, the interface controller can transmit them out the switch port back-to-back while complying only with the minimum 9.6 µs interpacket gap required by the 10 Mb/s IEEE 802.3 Standards.

Another feature of the PowerHub polling architecture is that it allows each software process to run to completion before the next process kicks in. The goal is to avoid the situation where the software and hardware architectures have to handle the extremely complex scenarios where two separate processes can inconsistently modify/manipulate the same piece of data at the same time.

Other switch architectures employ hardware or software locks to avoid the above problem (i.e., semaphores, locked bus cycles, or test-and-set instructions). Processing each lock can add significant processing overhead latency; for example, it can take as long as 3 µs to process a lock. In a system that is less optimized, it can take the processing of several locks per packet, further increasing the processing overhead latency per packet.

So to avoid these limitations in the traditional designs, the PowerHub architecture employs the polling architecture where the interface controller uses descriptor lists and the burst mode communication method as described above [FORESYSWP96]. In reality, these memory-based descriptor lists are realized as simple FIFO queues where the interface controller adds packet status items at one end and a processor removes these items at the other end.

6.4 PACKET PROCESSING IN THE POWERHUB

In the PowerHub, six separate processes service each packet forwarded by the system, with the processes communicating through FIFO descriptor queues. These FIFO descriptor queues are implemented in the shared memory as data structures that are accessed whenever necessary by PowerHub hardware processing elements. As many as 22 RISC processors can be supported within a PowerHub chassis. The PowerHub has additional processing capabilities in the network interface (NIM) controllers.

These NIM controller processing elements are implemented as microprogrammed engines that process FIFO descriptor queues and interact with the PowerHub packet engine's RISC processors. Figure 6.8 shows how the I/O Processor and Main CPU share processes when the system is using two of the Packet Engine's CPUs. Figure 6.9 illustrates how processes are shared among four CPUs when the Packet Accelerator is installed on the Packet Engine.

In addition to the hardware processes on the interface (NIM) controllers, the following six processes also service each packet forwarded through the system:

1. **Packet Reception:** The network interface controller (NIC) monitors its attached network for arriving packets and stores them in buffers specified by a receive FIFO descriptor queue.

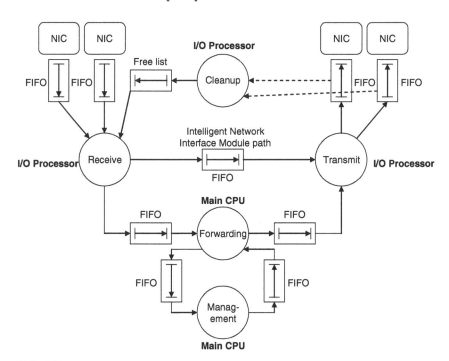

FIGURE 6.8   Illustrating how the I/O Processor and Main CPU share processes using two of the Packet Engine's CPUs.

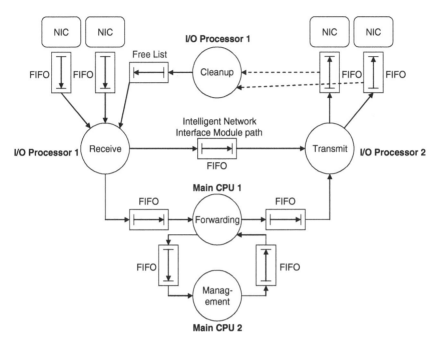

FIGURE 6.9   Illustrating how processes are shared among four CPUs when the Packet Accelerator is installed on the Packet Engine.

2. **Receive Polling:** The receive FIFO descriptor queues are then polled in a round-robin fashion by one or more I/O Processors, checking for incoming packets. Once a packet is received, it is checked by the I/O Processor for errors, which also updates the receive statistics. The packet's descriptor is then marked to indicate the packet is available for the next process on the forwarding path.

3. **Forward Polling:** The receive FIFO descriptor queues in turn are polled in a round-robin fashion by the Main CPU, checking for incoming packets. For each incoming packet, the CPU examines the packet header and determines the forwarding instructions according to the packet type and rules configured or set for that port (e.g., Layer 2 forwarding, Layer 3 forwarding, filtering, etc.).

   Special packets such as SNMP requests, RIP updates, and ICMP Pings may be directed to the PowerHub's packet engine (which holds also the route processor) where the Main CPU processes them. Once the forwarding decision is made, the receive FIFO descriptor for the packet is then updated with the destination port or ports (for multicast traffic). The destination is specified in the descriptor using a port mask, which is a bit map that specifies the destination port or ports to which the packet should be forwarded.

4. **Transmit Polling:** The receive FIFO descriptor queues are polled by the I/O Processor to check for packets marked with instructions for forwarding

outside the PowerHub. The I/O Processor copies a pointer to the packet's buffer to the transmit FIFO descriptor queue of each port specified in the port mask. Only the pointer is copied and not the packet itself.

If the transmit FIFO descriptor queue is not empty, a transmit-demand command is issued to the corresponding destination ports' network interface controllers. After forwarding a packet outside the PowerHub, the I/O Processor moves the pointer to the freed packet buffer from the "free" list to the receive FIFO descriptor queue.

5. **Packet Transmission:** Each transmit FIFO descriptor queue for packets to be transmitted is polled by its associated interface controller periodically. The transmit FIFO descriptor queue are also polled immediately after the interface controller receives a transmit-demand command.

Once a packet has been transmitted, the interface controller marks in the corresponding transmit FIFO descriptor (of the packet) a "completion status" indicator and error information related to the packet. The interface controller then immediately polls again the transmit FIFO descriptor queue for more work.

6. **Transmit Cleanup:** Each transmit FIFO descriptor queue is polled by the I/O Processor for "completion status," and the statistics for each transmitted packet are updated. As soon as the packet has been transmitted out all the destination ports on which it was queued, the I/O Processor reclaims the packet's buffer and returns it to the free list.

Each of the above software processes on the forwarding path is designed and optimized to continuously run in their various processors in a tight loop. The most computationally intense process is the Forward Polling (Step 3 above). This makes the performance of the Main CPU (where the Forward Polling is performed) the main limiting factor for the overall PowerHub packet forwarding capacity.

The remaining three software processes (Receive Polling, Transmit Polling, and Transmit Cleanup) are designed to be simple enough for the I/O Processor to multitask among them and still support the Main CPU's achievable packet forwarding rate.

6.5 LOOKING BEYOND THE FIRST-GENERATION ARCHITECTURES

Various switch and router architectures have emerged over the years of internetworking since 1990s. Each architecture has made its mark and contributed in various ways to the performance improvements seen in today's network devices and networks. A majority of the first-generation switches and routers were often designed based on shared bus and shared memory architectures with a single, centralized processor. However, as network sizes and traffic loads grew and the information in forwarding tables continued to increase, packet forwarding rates in switching and routing devices were also required to increase. Switch and router designs continued to evolve in architecture and in how the system divides up processor cycles among the various tasks running in the system. The architectures had to evolve to accommodate the added processing burden imposed by the growing network sizes and traffic loads. Shared-memory-based switch and router architectures had to be enhanced or redesigned all together, and newer and improved designs based on crossbar switch fabrics emerged.

These improvements delivered higher system reliability and packet forwarding capacities and made system performance (e.g., speed, latency, and data loss) much more predictable when operating under the larger and more dynamic traffic loads generated by newer end-user applications.

In addition to the use of higher capacity switch fabrics, one approach that has made a lasting impact on switching and routing system design and has significantly improved system performance is to logically decouple the packet forwarding functions (i.e., data or forwarding plane) from the control functions (i.e., control plane). The control functions run on the switching and routing system's centralized CPU or in CPU's located on distributed forwarding line cards in some switching and routing architectures.

This design approach was that the two planes are separated, allowing the two set of functions to run without interrupting or impeding each another. This allows the forwarding functions to reach their target forwarding speeds and avoid unpredictable latencies and lost data during packet forwarding. The control functions can also run in their isolated plane without impeding the forwarding functions.

To make a forwarding decision, a router must compare a packet's destination IP address with entries in a forwarding table the router maintains. The way in which the forwarding table is constructed and how its contents are searched affects both the system performance and packet forwarding rates. The first generation of switches and routers used software-based forwarding (on a centralized processor) to look up the forwarding information in the forwarding table for every packet forwarded (a method Cisco Systems calls process switching).

To improve the performance of the software-based forwarding (process switch-ing) systems, flow/route cache-based schemes were proposed. Techniques in this category are referred to by Cisco as fast switching and optimum switching. Cisco optimum switching differs from Cisco fast switching in that a router that uses optimum switching employs a more efficient tree structure in its forwarding table to enable it perform faster forwarding table lookups. Optimum switching exploits, in addition, the pipelining characteristics of the RISC processors (used on some Cisco routing platforms) to allow faster system performance.

These packet forwarding methods yield higher forwarding rates by allowing the system to forward a packet using a flow/route cache. The cache entries are created by the forwarding table lookup of the initial packet of a flow sent to a particular destination address. Destination address-to-exit port mappings are stored in the flow/route cache to allow the system to make faster forwarding information lookup and packet forwarding.

In the route cache-based systems, when a packet is copied to packet memory and the destination address is in the route cache, forwarding can be done much faster. This means that each time a packet is forwarded, it does not have to be processed using the normal forwarding table unless it is a first packet of a flow. The forwarding is done using the cache and the packet header is rewritten and sent to the outgoing port that leads to the destination. Subsequent packets of the same flow going to the same destination use the cache entry created by the first packet.

Latency and data loss have become even more important quality of service (QoS) metrics in today's networks that carry multimedia traffic. The high attention given to these metrics is due to the shift in traffic patterns in today's networks, a pattern characterized by a large number of real-time traffic (voice, video, interactive gaming, etc.) and short-lived flows (from TCP sources).

However, supporting a large number of short flows results in longer latencies in flow/cache-based switching and routing architectures, because the first packet in the flow must be processed and forwarded in software. Subsequent packets of the same flow would then be forwarded much faster using the flow cache (whose contents are created from the destination IP address-to-outgoing port mapping gleaned from the first packet forwarding process). Using this forwarding method, all routing devices on the best path to the destination must forward the first packet in the same fashion resulting in excessive delays even for a short flow.

Packet delay variation (PDV) was also another QoS metric of concern. Without decoupling the forwarding functions from the control functions on the shared CPU, variable packet processing and forwarding latencies can occur and can result in PDV. PDV is the biggest problem in applications such as voice, video, and interactive gaming. So, minimizing PDV became a significant gain that came with separating the forwarding and control functions within a device.

Switch and routing architectures evolved (and continue) to not only better address packet forwarding speeds, latency, and data loss but also to address newer and emerging concerns like system reliability and resiliency, scalability, enhanced network security with access control, efficient multicast support, energy consumption, and device footprint (office and storage space). The subsequent chapters will look at some newer architectures, starting with earlier generation ones.

7

ARCHITECTURES WITH BUS-BASED SWITCH FABRICS: CASE STUDY— CISCO CATALYST 6000 SERIES SWITCHES

7.1 INTRODUCTION

The Cisco Catalyst 6000 family of switch/routers consists of the Catalyst 6000 and 6500 Series that are designed to deliver high-speed Layer 2 and 3 forwarding solutions for service provider and enterprise networks. This chapter describes the different Catalyst 6000 series switch/router architectures [CISCCAT6000, MEN-JUS2003]. The Catalyst 6000 series supports several architectural options in terms of Layer 3 routing and forwarding capabilities.

Layer 3 routing and forwarding capabilities in the Catalyst 6000 and 6500 switches are handled by two important components: the Multilayer Switching Feature Card (MSFC) and the Policy Feature Card (PFC). These two key components, along with other functions required for system configuration and operation, are implemented on a specialized module in the Catalyst 6000/6500 called the Supervisor Engine.

The PFC provides the hardware-based forwarding engine functions required to perform in the Catalyst 6000/6500, Layer 2 and 3 forwarding, quality of service (QoS) classification, policing, shaping, priority queuing, and QoS and security access control list (ACL) processing. For Layer 3 forwarding, the PFC requires the services of a routing engine (or route processor) in the chassis to generate the necessary routing information needed for the flow/route cache or forwarding tables used by the Layer 3 forwarding engine.

Switch/Router Architectures: Shared-Bus and Shared-Memory Based Systems, First Edition. James Aweya.
© 2018 The Institute of Electrical and Electronics Engineers, Inc. Published 2018 by John Wiley & Sons, Inc.

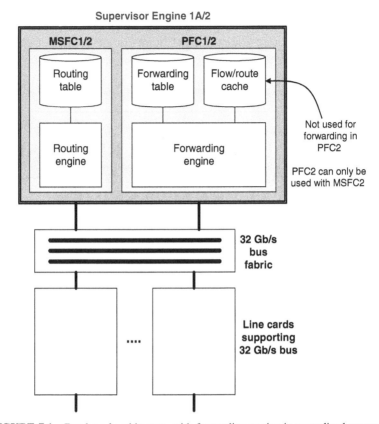

FIGURE 7.1   Bus-based architecture with forwarding engine in centralized processor.

The MSFC provides the control plane (routing engine) functions required by the PFC to perform Layer 3 forwarding. Based on the architecture categories defined in Chapter 3, the architectures discussed here fall under "Architectures with Bus-Based Switch Fabrics and Centralized Forwarding Engines" (see Figure 7.1)

7.2 MAIN ARCHITECTURAL FEATURES OF THE CATALYST 6000 SERIES

This section highlights the main architectural features of the Catalyst 6000 series of switch/routers. Compared to the Catalyst 6500 Series, the Catalyst 6000 Series offers lesser capabilities in terms of packet forwarding performance and scalability. The Catalyst 6000 Series offers a more cost-effective solution for enterprises and service providers not requiring the higher performing Catalyst 6500 Series.

The Catalyst 6000 Series has a backplane bandwidth of 32 Gb/s (using a shared switching bus architecture) and Layer 2/3 forwarding capacity of up to 15 million

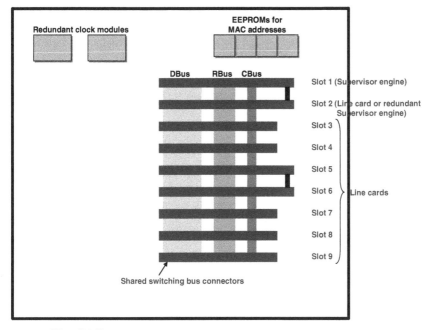

DBus = Data Bus
RBus = Results Bus
CBus = Control Bus also referred to as Ethernet out of band channel (EOBC)

FIGURE 7.2   Catalyst 6009 backplane.

packets-per-second (Mpps). The Catalyst 6500 Series architecture, on the other hand, supports a backplane bandwidth of up to 256 Gb/s and Layer 2/3 forwarding in excess of 200 Mpps.

The Catalyst 6000 Series supports different chassis options, a six-slot chassis in the Catalyst 6006 and nine-slot chassis in the Catalyst 6009. These switch/routers support a wide range of network interface types and port densities: 384 10/100BASE-T/TX Ethernet ports, 192 100BASE-FX Ethernet ports, and up to 130GbE (Gigabit Ethernet) ports (in the nine-slot chassis). The main identifying components on the Catalyst 6006 and 6009 backplane (Figure 7.2) are the following:

- Slot 1 supports the Supervisor Engine.
- Slot 2 supports either a line card or a redundant Supervisor Engine.
- Slots 3–9 support line cards.
- Supports bus connectors for the 32 Gb/s shared switching bus: Data Bus (DBus); Results Bus (RBus); Control Bus (CBus) (or Ethernet out-of-band channel (EOBC))
- Clock module with redundancy (primary and secondary).
- Ethernet MAC address EEPROMs.

The line cards used in the Catalyst 6000 Series are referred to as the "nonfabric-enabled" or "Classic" line cards. These line cards can be used in both the Catalyst 6000 and Catalyst 6500 Series and connect to the 32 Gb/s shared switching bus. They can also be used in all Catalyst 6000 and Catalyst 6500 Series chassis types as long as they support the 32 Gb/s switching bus.

Other line card types referred to as "fabric-enabled" and "fabric-only" are available for the Catalyst 6500. These line card types support connectivity to a crossbar switch fabric.

7.3 HIGH-LEVEL ARCHITECTURE OF THE CATALYST 6000

Both Cisco Catalyst 6000 and 6500 Series support a common 32 Gb/s shared switching bus architecture as shown in Figure 7.3. The Catalyst 6000, in particular, supports only the 32 Gb/s shared switching bus as the backplane connectors in Figure 7.2 illustrate. The Catalyst 6500, on the other hand, supports both the 32 Gb/s shared switching bus (same architecture as in the Catalyst 6000 Series) and an option for a stand-alone 256 Gb/s Switch Fabric Module (SFM) or an integrated 720 Gb/s crossbar switch fabric on Supervisor Engine 720. The exception is the Catalyst 6500 with Supervisor Engine 32 that supports only a 32 Gb/s shared switching bus.

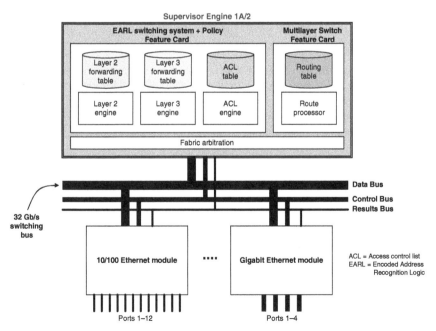

FIGURE 7.3 High-level view of the Cisco Catalyst 6000 and Catalyst 6500 switching bus architecture.

The 32 Gb/s backplane is designed as an advanced pipelining shared switching bus and consists of three separate sub-buses: the Data Bus, the Results Bus, and the Control Bus (which is also called the Ethernet out-of-band channel). The Classic (or nonfabric-enabled) line card types connect to the 32 Gb/s shared switching bus via the bus connectors on the backplane shown in Figure 7.2.

The DBus has a bandwidth of 32 Gb/s and is the main bus that carries data from one system module or line card to another. The RBus carries forwarding information (obtained after forwarding table lookup) from the forwarding engine located on the Supervisor Engine back to all the line cards. The CBus or EOBC carries control and management information between the line card port ASICs and the network management entity in the Supervisor Engine.

The 32 Gb/s pipelining switching bus is designed as a shared medium bus that allows all the frames transmitted on it to be visible to all the modules and ports attached to the bus. The pipelining mechanism working with the shared switching bus allows efficient reception of a transmitted frame after a forwarding decision is made by the forwarding engine. It also allows the flushing of a transmitted frame at the nondestination ports once the destination ports have received the frame (i.e., they are instructed to ignore the frame).

7.3.1   32 Gb/s Switching Bus Operating Modes

The 32 Gb/s switching bus has two operational modes: pipelining and burst modes. These two modes are described in the following sections.

7.3.1.1   Pipelining Mode   In the typical or conventional shared medium bus architecture, only a single frame can be transmitted (i.e., propagated) on the bus at any given time. In such an architecture, if the frame is transferred across the shared switching bus from a port (to the forwarding engine) before the forwarding table lookup is completed, the shared switching bus stays idle until the address lookup is completed. Pipelining allows the ports to transmit multiple frames back-to-back on the switching bus while waiting for the results of the first frame sent to be received from the forwarding engine.

With pipelining, ports are allowed to transmit multiple frames on the shared switching bus before the result of the first frame address lookup is obtained. Immediately after transmitting the first frame, the second frame (which can be from any port) is transmitted across the shared switching bus and pipelined for forwarding table lookup at the forwarding engine. The system allows the address lookup process at the forwarding engine to occur in parallel to the transfer of the multiple frames across the shared switching bus.

In the pipelining mechanism, the 32 Gb/s switching bus allows 31 frames to be transferred back-to-back across the shared switching bus (and pipelined at the forwarding engine for address lookup operation) before the result of the first frame is received. The 32nd frame to be sent (after the sequence of transmitted 31 frames) must wait until the pipelining process once again allows it to be transmitted on the shared switching bus as the first frame of a new sequence of 31 frames.

7.3.1.2 Burst Mode   The burst mode feature together with a transmit threshold mechanism allows fair allocation of the 32 Gb/s shared switching bus bandwidth to the ports. To understand how this works, let us consider the following case. If a port transmits just a single frame each time it is permitted to transmit on the shared switching bus, depending on the lengths of the frames transmitted by all ports, there could potentially be an unfair bandwidth allocation to some ports during heavy traffic load conditions.

Let us assume, for example, that two ports (Ports A and B) have data to send and with Port A having 150 byte frames while Port B 1500 byte frames. With a simple allocation policy without thresholds, Port B with the 1500 byte frames can send 10 times more data than Port A with 150 byte frames. This unfairness arises because the ports alternate in the arbitration process for the shared switching bus and when granted access each port transmits just one frame at a time. The burst mode feature enables a port to transmit multiple frames on the shared switching bus but subject to a threshold.

With the burst mode feature, a port can transmit multiple frames on the shared switching bus but the amount of bandwidth it consumes is controlled independent of the frame size it transmits. A count of the number of bytes a port has transmitted is maintained by its port ASIC (in a counter) that is compared to a threshold. The threshold values are computed by the system to ensure fair distribution of the shared switching bus bandwidth.

As long as the port byte count is below the threshold value, the port is permitted to transmit more frames if it has any. When the byte count goes above the threshold, the port is not permitted to send additional frames after transmitting the current frame and stops further frame transmission. This is because the fabric arbitration logic (see Figure 7.3) upon sensing this threshold exceeding condition removes bus access for the port in question.

7.3.2   Control Plane and Data Plane Functions in the Catalyst 6000

Similar to routers, routing and switching in multilayer switches rely on two key functions commonly referred to as the control plane and data plane. The control plane (realized through a route processor, also called the control engine) is responsible for running the routing protocols, generating and maintaining the routing table, and maintaining all of the management functions of the switch including device security and access control.

The data plane (realized through the forwarding engine(s)) is responsible for forwarding a packet on its way to the destination using the routing information generated by the control plane.

- **The MSFC as the Routing Engine:** The Catalyst 6000 uses a centralized control plane functionality that resides in a daughter card module called the

MSFC, which is one of two key components on the Supervisor Engine. In Supervisor Engines 1A and 2, the MSFC runs the routing protocols and maintains the routing table. It communicates with the hardware forwarding engines in the system across an out-of-band bus called the CBus (or EOBC).

- **The EARL (Encoded Address Recognition Logic) as a Layer 2 Forwarding Engine:** The EARL is a centralized Layer 2 processing and forwarding engine (in the Catalyst 6000 and 6500 Supervisor Engines) for learning MAC address locations of connected stations and forwarding packets based upon the learned MAC addresses. The EARL maintains the VLAN, MAC address, and port relationships in a Layer 2 forwarding table as illustrated in Figure 7.3. These relationships are used to make Layer 2 forwarding decisions in hardware using Layer 2 forwarding ASICs. In some Supervisor Engine architectures as discussed below, the EARL functions (i.e., Layer 2 forwarding functions) are fully integrated into the PFC.

- **The PFC as the Forwarding Engine:** The PFC implements the packet forwarding engine in the Catalyst 6000. It supports the forwarding engine ASICs that enable packet forwarding at data rates of up to 15 Mpps in the Catalyst 6000. The PFC also provides Layer 3/4 level packet field inspection and processing, allowing some security and QoS features to be supported based upon the Layer 3 and Layer 4 parameters of user packet traffic. As discussed below, the PFC supports Layer 3 forwarding only with the addition of an MSFC (which supports the Layer 3 control plane functions) in the system. The PFC can be installed and used just by itself for Layer 2 forwarding and simple Layer 3/4 packet inspection/processing, without an MSFC installed in the system.

The MSFC is essentially an IP router on a daughter card installed in the system, providing full Layer 3 routing functionality and enabling the Catalyst 6000 to perform Layer 3 forwarding. In a Layer 3 forwarding configuration, the MSFC provides the control plane component of Layer 3 forwarding engine (i.e., populating and maintaining the routing table).

The PFC provides the data plane component of Layer 3 forwarding engine (i.e., forwarding table lookups, rewriting frame and packet headers, and forwarding packets undergoing routing to the appropriate egress port). This means an MSFC must be installed with a PFC for full Layer 3 control and data plane operations to take place.

7.4 CATALYST 6000 CONTROL PLANE IMPLEMENTATION AND FORWARDING ENGINES: SUPERVISOR ENGINES

The Catalyst 6000 Series supports two versions of the Supervisor Engine – Supervisor Engine 1A and Supervisor Engine 2. Supervisor Engine 1A is the first version of the integrated routing and forwarding engine designed for the Catalyst

6000 family. A Supervisor Engine must be installed in a Catalyst 6000 for it to function since all the required "intelligence" for system operation resides in this module. As shown in Figure 7.2, the Supervisor Engine sits in Slot 1 in the chassis. Slot 2 can accommodate a secondary redundant Supervisor Engine when required, or otherwise can be used for a line card.

For both Catalyst 6000 and 6500, one Supervisor Engine is sufficient for system operation, and in a redundant configuration, only one Supervisor Engine of the two engines needs to be active at one time. However, in the redundant configuration, both Supervisor Engines maintain the same state information, including Layer 3 routing information and tables, Layer 2 information including Spanning-Tree topology and Layer 2 forwarding tables, and system management information. In this configuration, if the primary Supervisor Engine fails, the redundant Supervisor Engine takes over without noticeable interruption in system operation.

7.4.1 Supervisor Engine 1A Architecture

Supervisor Engine 1A supports a flow/route cache-based forwarding scheme and provides forwarding performance of 15 Mpps. The Supervisor Engine 1A is targeted for deployment in network access layer scenarios such as in a wiring closet of an enterprise network, a server farm, and the main distribution frame (MDF) of an office building in an enterprise network.

As illustrated in Figures 7.3 and 7.4, Supervisor Engine 1A has the following main components: the EARL switching system, MSFC, and PFC. Each component supports a number of critical functions necessary for system operation. Supervisor Engine 1A is designed to support three different configurations as described in the following sections.

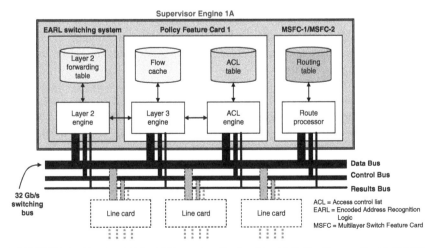

FIGURE 7.4   Catalyst 6000 with Supervisor Engine 1A with Policy Feature Card 1 (PFC1) and Multilayer Switch Feature Card (MSFC).

7.4.1.1 Supervisor Engine 1A with Only an EARL Switching System In this configuration, Supervisor Engine 1A has only an EARL switching system and with no MSFC and PFC present in the system. This is the most basic configuration allowing only basic Layer 2 forwarding based on MAC addresses.

The system does not support any Layer 3-based QoS or security ACLs, but only port-based and destination MAC address-based class of service (CoS). In this configuration, the Catalyst 6000 is reduced to a simple Layer 2 switch with no support for Layer 3 forwarding, QoS, and security classification capabilities. In this basic mode, the Supervisor Engine 1A can provide only Layer 2 forwarding of up to 15 Mpps.

7.4.1.2 Supervisor Engine 1A with an EARL Switching System Plus PFC1 In this configuration, Supervisor Engine 1A has an EARL switching system in addition to a PFC but no MSFC is present in the system. This is like removing the MSFC block from Figure 7.4 leaving only the EARL and PFC. This configuration provides Layer 2 forwarding with Layer 3 QoS and security ACLs services only. Layer 3 forwarding and routing is not supported. The system supports Layer 2 forwarding, Layer 3 QoS classification, queuing, and security filtering at data rates of 15 Mpps. These Layer 2 and 3 services are supported even though Layer 3 forwarding and route processing are not performed (unless an MSFC is added to provide route processor functions).

In this configuration, the Supervisor Engine 1A has the basic Layer 2 forwarding engine that inspects the local Layer 2 forwarding table to determine the egress port, and possibly VLAN, for Layer 2 forwarded packets. The PFC implements a Layer 3 forwarding engine, route/flow cache, and an ACL processing engine with a local ACL table. The PFC does not perform Layer 3 forwarding because no MSFC is present to provide the route processing functions required to generate the required Layer 3 routes and next hop information. The PFC, however, can perform Layer 3/4 QoS classification and ACL filtering using its local ACL engine and ACL table.

The local ACL table is maintained in ternary content addressable memory (TCAM). The TCAM stores and maintains the ACL information in a data format (i.e., data structures) that can be easily inspected by the ACL engine. A number of tasks are processed in parallel when a packet arrives at Supervisor Engine that requires ACL filtering. While the Layer 2 forwarding engine determines the egress port and VLAN for the packet by examining the forwarding information in the Layer 2 forwarding table, the ACL engine inspects its ACL table to determine if the packet is to be permitted or denied into the system. The lookups in Layer 2 forwarding table and the ACL table are performed in parallel, thus preventing QoS classification and ACL processing of traffic to not adversely affect the 15 Mpps forwarding rate of the switch.

7.4.1.3 Supervisor Engine 1A with an EARL Switching System Plus PFC1/ MSFC1 (or 2) This configuration allows for full Layer 3 routing and forwarding in the Catalyst 6000 using the Supervisor Engine 1A. Here, the Supervisor Engine 1A module has a PFC1 and MSFC1 or MSFC2 in the system (Figure 7.4). This configuration provides Layer 2 forwarding in addition to full Layer 3 routing and forwarding with the corresponding Layer 3 QoS and security services. This configuration enables the Catalyst 6000 to forward IP traffic at 15 Mpps.

The MSFC1 and MSFC2 share a similar architecture with the main difference being only in packet forwarding performance. The MSFC1 is designed with an R5000 200 MHz processor, up to 128 MB memory, and packet forwarding rate of up to 170 kpps in software. The MSFC2 supports an R7000 300 MHz processor, up to 512 MB memory, and packet forwarding rate of up to 650 kpps in software. They both can support Layer 3 packet forwarding in hardware at 15 Mpps in the Supervisor Engine 1A.

In this configuration, the Layer 3 forwarding engine on the PFC1 can perform Layer 3 forwarding, because route processing can now be done using the MSFC. The presence of the MSFC allows the Layer 3 forwarding engine to forward packets requiring routing such as is required in inter-VLAN communications. This Layer 3 forwarding can be done in addition to the other PFC features, such as QoS classification and ACL filtering. The Supervisor Engine 1A with PFC1 and MSFC1/MSFC2 employs a route/flow cache to forward Layer 3 traffic.

The flow cache (which is maintained on the PFC1 (Figure 7.4)) is used to forward Layer 3 packet flows through the Catalyst 6000. The first packet in a flow is always sent to the MSFC, which examines the local forwarding table it maintains to determine the next hop information for this first packet. The MSFC makes a Layer 3 forwarding decision and forwards the packet and the forwarding instructions back to the Layer 3 forwarding engine in the PFC1.

The Layer 3 forwarding engine then extracts the packet forwarding information and writes this information into its flow cache. When subsequent packets (belonging to the same flow as the first packet) are received and match the flow cache entries in the PFC1, they are Layer 3 forwarded directly by the PFC1 Layer 3 forwarding engine, rather than sent to the MSFC for processing and forwarding.

The main limitation of the flow cache-based method for Layer 3 forwarding is that the initial Layer 3 forwarding table lookup is performed by the MSFC software process. The first packet in a Layer 3 flow must be sent to the MSFC for Layer 3 forwarding table lookup and forwarding. This means that in a network environment that has many short-term Layer 3 flows being set up at the same time, the MSFC software process can easily be overwhelmed by the many flows it has to handle. This problem becomes more acute particularly in enterprise and service provider core network environments, where many short-term connections can be established at the same time.

7.4.1.4 Details of Packet Processing in the Supervisor Engine 1A

The EARL module has its own local processor (for Layer 2 forwarding), which is referred to as the switch processor. The switch processor is responsible for running the Layer 2 protocols of the switch (e.g., Spanning Tree Protocol (STP), IEEE 802.1AB Link Layer Discovery Protocol (LLDP), and VLAN Trunking Protocol (VTP)), as well as implementing some QoS and security related services necessary for the PFC

Layer 3/4 data plane operations. The MSFC also has its own local processor, which simply can be referred to as the route processor and is responsible for implementing the Layer 3 control plane functions.

On Supervisor Engine 1A, the first packet in a flow that does not have an entry registered in the flow cache in the PFC1 is sent to the MSFC for software-based forwarding. The MSFC extracts the packet's destination IP address and performs a lookup in its local forwarding table to determine how to forward the packet. After the MSFC software process has forwarded the first packet of the new flow, the PFC1 receives and uses this forwarding information to program its flow cache so that it can forward subsequent packets in the same flow directly without MSFC intervention.

The forwarding decisions in the Supervisor Engine 1A are handled by three components (Figure 7.4): EARL switching system for Layer 2 MAC address-based forwarding, MSFC for Layer 3 forwarding of the first packet in a flow, and the PFC for Layer 3 forwarding of subsequent packets in a flow and ACLs processing (for implementing QoS and security services).

The EARL switching system (Layer 2 ASIC) learns MAC addresses within a broadcast domain (or VLAN) to create a Layer 2 forwarding table, which in turn is used to forward packets at Layer 2. The EARL module also identifies which packets (within a flow) need to be forwarded at Layer 3– Packets sent to destinations outside the broadcast domain have to be Layer 3 forwarded.

After forwarding the first packet, the MSFC generates an entry to be installed in the flow cache (in the PFC), which the Layer 3 forwarding engine (in the PFC) uses to forward subsequent packets in the flow in hardware. To facilitate the forwarding of packets requiring routing, the MSFC (considered the default gateway for routing traffic) registers its assigned MAC address with the Layer 2 forwarding engine so that upon examination of a packet, it can decide if the packet is to be sent to the MSFC or not.

The Layer 2 forwarding engine forwards packets requiring routing to the Layer 3 forwarding engine first, which in turn may forward them to the MSFC for further processing. Traffic to the MSFC (i.e., the default gateway for packets going to another VLAN) is sent to its known and registered MSFC MAC address at the Layer 2 forwarding engine.

After the Layer 2 forwarding engine determines that Layer 3 forwarding needs to take place (i.e., traffic to outside the broadcast domain), the services of the Layer 3 forwarding engine are engaged. The Supervisor Engine 1A uses flow-caching where a flow can be defined as a traffic stream from a specified source IP address to a specified destination IP address. The flow cache can also store in addition Transmission Control Protocol (TCP) and User Datagram Protocol (UDP) port numbers as part of the flow cache entry.

When the first packet in a flow arrives at the PFC, a lookup is performed in the flow cache by the Layer 3 forwarding engine to determine if an entry for the packet exists. If the cache contains no related entry, the packet is sent to the MSFC, which uses the destination IP address to perform a lookup in the forwarding table it maintains (to determine the next hop IP address, the egress port, and destination

VLAN). The packet is then forwarded to the next hop and the forwarding information is used to create an entry in the PFC flow cache. Subsequent packets in the same flow can then be forwarded by the PFC hardware (by the Layer 3 forwarding engine using the newly created flow cache entry).

A maximum of 128,000 entries can be supported by the flow cache and there are three options for creating flow cache entries:

- **Destination-Only Flow Option:** In this configuration, a flow entry is created in the flow cache based on the destination IP address. This option utilizes less PFC flow cache space as multiple source IP addresses communicating with one destination IP address (e.g., a server) result in only a single-flow cache entry.
- **Source-Destination Flow Option:** In this configuration, a flow entry is created based on both the source and destination IP addresses. This option consumes more entries in the flow cache; for example, if four source IP addresses are communicating with one destination IP address, then four entries are created in the flow cache.
- **Full-Flow Option**: This is a more resource-consuming option of using the flow cache because a flow entry is created not only from the source and destination IP addresses but also from the UDP or TCP port numbers.

To facilitate searches, the flow cache is split into eight pages of memory with each storage page capable of handling 16,000 entries. The PFC then uses a hashing algorithm to carry out lookups in the flow cache. The hashing algorithm has to be efficiently implemented since it is critical to how entries can be stored and lookups can be performed at high speeds.

Hashing algorithms work in a statistical manner that can lead to hash collisions. A hash collision occurs when the lookups for two packets with different parameters hash to the same location in memory. To account for such collisions, the Layer 3 forwarding engine (in the PFC) moves to the next page in memory to check if that location is used. This next page search process continues until either the lookup information is stored or until the eighth page is reached. If the lookup information still cannot be stored after the eighth page, then the packet is flooded out the switch (at Layer 2) or sent to the MSFC (for further (and possibly, Layer 3) processing).

Supervisor Engines 1A and 2, and the Distributed Forwarding Cards (DFCs) in the Catalyst 6500 all use TCAMs for storing and processing QoS and security ACLs. A TCAM has the capacity of 64,000 entries and is split into four main blocks. Two blocks are used for QoS ACLs input checking and output checking, and two blocks for security ACLs input and output checking.

The PFC1 and PFC2 have similar TCAM implementation and behavior. The implementations allow ACL lookups to take place at packet rates of up to 15 Mpps on Supervisor Engine 1A and 30 Mpps on the DFC and Supervisor Engine 2. ACL lookups are performed in parallel to the Layer 2 and 3 lookups, which results in no performance degradation when processing QoS or security ACLs.

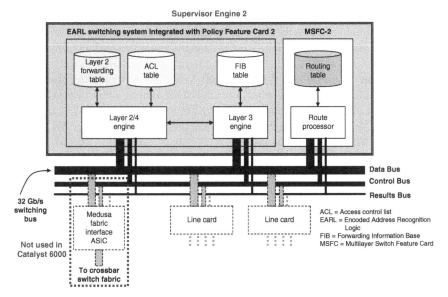

FIGURE 7.5 Catalyst 6000 with Supervisor Engine 2 with Policy Feature Card 2 (PFC2) and Multilayer Switch Feature Card 2 (MSFC2).

7.4.2 Supervisor Engine 2 Architecture

Supervisor Engine 2 provides higher packet forwarding performance and system resiliency compared to Supervisor Engine 1A. Supervisor Engine 2 (Figure 7.5) provides forwarding capacity of up to 30 Mpps when deployed in a Catalyst 6500 chassis using both fabric-enabled line cards and the Crossbar Switch Fabric Module (SFM).

The fabric-enabled line cards and SFM can only be used in the Catalyst 6500 Series and not in the Catalyst 6000. Furthermore, the SFM requires the use of Supervisor Engine 2 or higher, but Supervisor Engine 2 can still operate without a SFM in the system, that is, when used in the Catalyst 6000 that does not support the SFM. The backplane architecture in Figure 7.2 shows no support of a SFM.

Supervisor Engine 2 is more suitable for deployment in the core of service provider and large enterprise networks. The major difference between Supervisor Engine 1A and Supervisor Engine 2 is that Supervisor Engine 2 supports topology-based forwarding tables with highly optimized lookup algorithms (also referred to as Cisco Express Forwarding (CEF)) implemented in hardware. We have already seen above that Supervisor Engine 1A supports only flow cache-based forwarding. The forwarding tables in Supervisor Engine 2 can also be distributed to the line cards if they support local forwarding engines – distributed forwarding.

As explained in Chapter 1, distributed forwarding is a forwarding method that is based on distributing the forwarding tables (created from the topology of the network rather than from traffic flow caching) to the line cards so that forwarding can be done locally there. Supervisor Engine 2 comes in two configurations: Supervisor Engine 2 with EARL switching system integrated with PFC2, and Supervisor Engine 2 with EARL switching system integrated with PFC2/MSFC2.

7.4.2.1 Supervisor Engine 2 with EARL Switching System Integrated with PFC2   This configuration provides only Layer 2 forwarding with Layer 3 QoS and security ACLs and services, in addition to Private Virtual LAN (PVLAN) services. The Supervisor Engine 2 in this configuration has only a PFC2 installed (Figure 7.5, without any MSFC block). The PFC2 and PFC1 have similar functions (including Layer 3/4 QoS classification and security ACL filtering), although the PFC2 is two times faster than the PFC1 and can store more QoS and security ACLs in hardware.

With switch fabric-enabled line cards and a SFM installed in a system, the Supervisor Engine 2 with PFC2 is capable of forwarding packets and performing Layer 3/4 QoS classification and ACL filtering at speeds of up to 30 Mpps. However, with no MSFC present in this configuration to provide routing information, Layer 3 forwarding (routing) cannot be done.

As shown in Figure 7.5, the EARL switching system is actually integrated into the PFC2. The Layer 2 and ACL engine are combined to obtain a single Layer 2/Layer 4 engine. The capabilities of the Layer 2 forwarding engine are enhanced to include Layer 3/4 QoS classification and ACL filtering. The Layer 3 forwarding engine is not used for Layer 3 forwarding, because an MSFC2 is not present to generate the routing information required to populate the forwarding table.

7.4.2.2 Supervisor Engine 2 with EARL Switching System Integrated with PFC2/MSFC2   In this configuration, an MSFC2 is added to enable Layer 3 forwarding on the Supervisor Engine 2 with PFC2. Supervisor Engine 2 does not support the MSFC1. The Layer 3 forwarding engine on the PFC2 can now perform Layer 3 forwarding because routing information is now provided by the MSFC2 (Figure 7.5). This configuration enables Layer 2 forwarding with full Layer 3 routing and forwarding on the Catalyst 6000 and Catalyst 6500.

The addition of the MSFC allows the Layer 3 forwarding engine on the PFC2 to Layer 3 forward packets (e.g., inter-VLAN traffic) while also supporting all other features of the PFC, such as QoS classification and ACL filtering. The PFC2 and MSFC2 both use topology-based forwarding tables with optimized lookup mechanisms (Cisco Express Forwarding (CEF)). The MSFC2 is responsible for running the routing protocols, building the routing tables, and generating the appropriate CEF tables (which include the Forwarding Information Base (FIB) table and adjacency table) to be used by the PFC.

In this configuration, as soon as packets need to be Layer 3 forwarded, the Layer 3 forwarding engine in the PFC already has the necessary information in its forwarding table to forward the packet to the next hop, without having to send the first in a flow to the MSFC. This forwarding architecture avoids the problems associated with flow cache-based forwarding when operating in an environment that has a high number of short flow connections being established in very short time intervals.

7.4.2.3 Details of Packet Processing in the Supervisor Engine 2   The MSFC2 in Supervisor Engine 2 does not forward IP packets (apart from exception packets directed to it by the PFC2). Instead, the MSFC2 constructs and maintains the main copy of a distributed forwarding table (also known as a Forwarding Information Base). The FIB contains the most important information required for packet forwarding and this information is distilled from the routing table created by the routing protocols running in the MSFC2. The MSFC2 copies the FIB it generates directly to the forwarding hardware in the PFC2 so that all packets are forwarded in the PFC2 hardware and not by the MSFC2 software process.

It is important to note that a flow cache is also generated in the PFC2, but this flow cache is used for statistics collection (e.g., as in NetFlow) and not for Layer 3 forwarding of packets.

7.4.3 Multilayer Switching Highlights in Cisco Catalyst Switches

This section summarizes the main features of multilayer switching in the Catalyst 6000/6500 switches and other Cisco Catalyst switches to be discussed in later chapters. The discussion here sets the context for the forwarding methods used here and in the other architectures. Cisco Catalyst switches support two methods of hardware-based Layer 3 forwarding, some aspects of which have already been described above for Supervisor Engines 1A and 2. The methods differ in how the data plane components of Layer 3 forwarding can get the necessary control plane information required to forward packets.

7.4.3.1 Front-End Processor Approach with Flow-Based Forwarding   This method (called Multilayer Switching (MLS) by Cisco) represents the first method of hardware-based Layer 3 forwarding used by Cisco Catalyst switches. The method uses a flow-based model to populate a flow cache that includes the necessary control plane information required for the data plane to Layer 3 forward a packet. A flow simply represents a stream of IP packets, each sharing a number of identical parameters, such as the same source/destination IP address, or same source/destination TCP port or a combination of these.

An MLS Route Processor (MLS-RP) (i.e., the MSFC) provides control plane operations, while an MLS Switching Engine (MLS-SE) (i.e., the PFC) provides data plane operation. MLS requires that the first packet of a new flow (i.e., candidate packet) received by the MLS-SE be forwarded to the MLS-RP. The MLS-RP then makes a Layer 3 forwarding decision using a software process operating as part of its control plane and forwards the packet in software to its next hop but the packet exits the switch via the MLS-SE.

The MLS-SE receives the Layer 3 forwarding instructions in the returned processed (first) packet (i.e., enabler packet) from the MLS-RP that is on its way to the next hop. The MLS-SE then populates the flow cache with the forwarding information required to Layer 3 forward subsequent packets that belong to the flow associated with the first packet. Subsequent packets received by the MLS-SE can then be Layer 3 forwarded in MLS-SE hardware without requiring the packets to be sent to the MLS-RP because the flow cache now has the required forwarding information.

7.4.3.2 Distributed Forwarding Approach (aka Cisco Express Forwarding) This forwarding method, which uses optimized lookup algorithms and network topology-based distributed forwarding tables (method referred to by Cisco as CEF), is the newer generation of hardware-based Layer 3 forwarding. This is the preferred forwarding method used by modern-day Cisco and other high-performance switch/ routers and routers.

In the CEF architecture, the forwarding (or CEF) table is prepopulated with all the necessary Layer 3 forwarding information (distilled from the routing table). This allows the Layer 3 forwarding engine ASIC to forward all IP packets in hardware, unlike in the MLS approach that requires the first packet of a flow to be forwarded in software by the MLS-RP.

The CEF architecture is more efficient and scalable and avoids the performance limitations of MLS method in environments where thousands of new short flows are established in very short time intervals. The CEF architecture is very scalable because the main CEF table information can be distributed to multiple Layer 3 forwarding engines. This means that a switch/router or router can perform multiple Layer 3 forwarding operations simultaneously, one per CEF table and forwarding engine.

The route processor (control plane) component of switch/router or router is responsible for generating the information in the CEF table and updating it as network routing topology changes occur. The CEF table can be viewed as consisting of two tables: the Layer 3 forwarding table and the adjacency table that hold the Layer 2 addresses of the next hops and directly attached hosts.

7.5 CATALYST 6000 LINE CARD ARCHITECTURES

The Catalyst 6000 and 6500 employ two types of port ASICs for network connectivity. The PINNACLE ASIC is designed for Gigabit Ethernet network ports (Figure 7.6), and the COIL ASIC for 10/100 Mb/s Ethernet ports (Figure 7.7). These port ASICs provide connectivity from a network to the 32 Gb/s main shared switching bus or the 16 Gb/s local bus supported on the fabric-enabled and fabric-only line cards.

The port ASICs also support the Catalyst 6000 and 6500 congestion management mechanisms. The PFC on the Supervisor Engine (or the DFC on fabric-enabled line cards) is responsible for instructing these port ASICs on how a packet should be classified and queued for QoS processing.

As illustrated in Figure 7.6, each PINNACLE ASIC supports four Gigabit Ethernet ports and provides congestion management with per port buffering. For

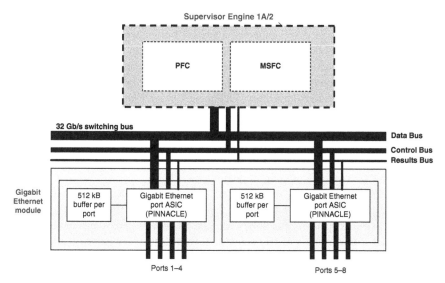

FIGURE 7.6   PINNACLE Gigabit Ethernet (GbE) Port ASIC and Buffering.

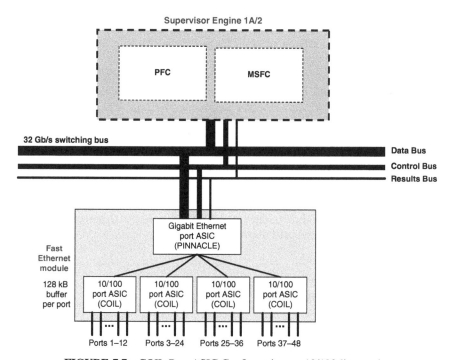

FIGURE 7.7   COIL Port ASIC Configuration on 10/100 line cards.

example, each 16-port Gigabit Ethernet line card holds four PINNACLE ASICs with each PINNACLE ASIC allocated a 512 kB buffer per port.

To prevent head-of-line (HOL) blocking to the shared switching bus fabric, a smaller amount of buffer is allocated to the receive (RX) queue (which is the queue that accepts frames coming from the network into the switch). A larger share of the buffering is allocated to the transmit (TX) queue (which is the queue that transmits frames from the switch to the network).

The allocation is done such that the ratio between the amount of buffers assigned to the transmit queue and receive queues is 7:1, which results in 448 kB of TX queue buffer and 64 kB of RX queue buffer. This buffer allocation strategy makes the Catalyst 6000 essentially an output queuing switch.

The PINNACLE ASIC handles QoS processing by assigning each port two RX queues and three TX queues. One queue out of the three TX queues is served in a strict priority fashion, while the other two queues are served using a weighted round-robin (WRR) scheduler. With the strict priority scheduler, the strict priority queue is allocated a fixed and guaranteed amount of bandwidth that is configured at the output port scheduling logic located in the port ASIC.

With the WRR scheduler, the two remaining queues are given scheduling weights that are relative to each other (i.e., sum of the normalized weights is equal to one) and where each queue is given bandwidth at the outgoing port proportional to its weight. If the three TX queues are configured as strict priority, high-priority, and low-priority queues, then the default port bandwidth allocation configuration will have 15% for strict priority queue, 15% for high-priority queue, and 70% for low-priority queue.

As Figure 7.7 illustrates, each COIL ASIC has 12 10/100 Mb/s Ethernet ports. Each PINNACLE ASIC, in turn, supports four COIL ASICs resulting in 48 ports on the line card. The 10/100 Mb/s Ethernet ports using the COIL ASIC work with their attached PINNACLE ASIC to implement congestion management. These dual ASIC line cards rely on a combination of control mechanisms in the COIL and PINNACLE ASICs to carry out congestion management.

Similar to the Gigabit Ethernet line card described above, the COIL ASIC supports buffering on a per port basis with each 10/100 Mb/s Ethernet port in the system allocated 128 kB of buffer. This 128 kB buffer is in turn divided between the TX and RX queues in a 7:1 ratio. The smaller RX buffer is used to prevent any HOL blocking problems as discussed above.

7.6 PACKET FLOW IN THE CATALYST 6000 WITH CENTRALIZED FLOW CACHE-BASED FORWARDING

This section describes the flow of packets through the Catalyst 6000 (with the 32 Gb/s shared switching bus) and a centralized flow cache maintained by the Supervisor Engine 1A. The processing steps are described in Figures 7.8–7.10.

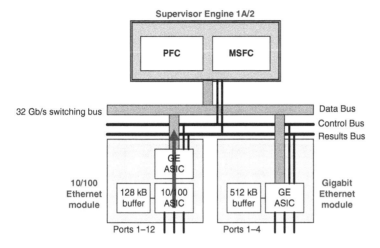

FIGURE 7.8 Step 1: Packet enters the switch from network.

Step 1 (Figure 7.8): Packet Enters an Input Port on the Switch from the Network

- A packet from the network enters an input port and is temporarily stored in the RX buffer. The packet is held in the RX buffer while the PINNACLE ASIC arbitrates for access to the 32 Gb/s shared switching bus.

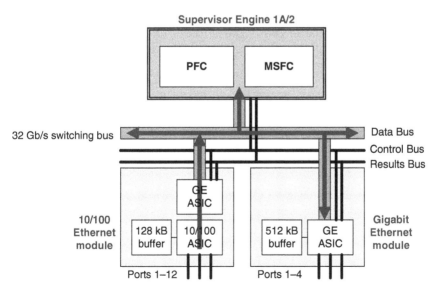

FIGURE 7.9 Step 2: Packet sent across 32 Gb/s switching bus and lookup takes place in Supervisor Engine.

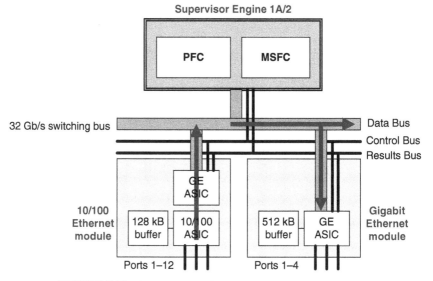

FIGURE 7.10   Step 3: Packet forwarded from switch to network.

- The 32 Gb/s shared switching bus is a shared medium allowing all the ports and modules connected to it to sense a transmitted packet as it propagates on the bus. Each line card has a local arbitration mechanism that allows each port on each PINNACLE ASIC to request for access to the shared switching bus.
- The local arbitration mechanism communicates with the central arbitration mechanism on the Supervisor Engine (see Figure 7.3), which then determines when each local arbitration mechanism is allowed to transmit packets on the shared switching bus.

Step 2 (Figure 7.9): Packet Sent Across the 32 Gb/s Switching Bus and Forwarding Table Lookup Takes Place in the Supervisor Engine

- Once the PINNACLE ASIC has been granted access to the shared switching bus by the central arbitration mechanism, the packet is transmitted across the bus.
- As the packet propagates along the shared switching bus, all the connected ports and modules start copying that packet into their TX buffers.
- The connected PFC, which also monitors the shared switching bus, senses the transmitted packet and initiates a forwarding table lookup process. First, the PFC references its Layer 2 forwarding table to determine if Layer 2 forwarding is required. If the packet is destined to a station located in the same VLAN as the source station served by the switch, then Layer 2 forwarding is carried out.

- However, if the Layer 2 destination address in the packet is the MSFC's registered MAC address, then the Layer 3 forwarding Engine in the PFC checks its flow cache to determine if a forwarding entry exists for the packet. If no flow entry exists, the packet is sent to the MSFC for further processing. But if a flow entry does exist in the flow cache, the PFC uses the packet's destination IP address to perform a lookup in the flow cache for the next hop MAC address, outbound port, and VLAN associated with the packet's destination.

Step 3 (Figure 7.10): Packet Forwarded from the Switch Through the Outbound Port to the Network

- After the lookup process above, the Supervisor Engine identifies the outgoing port for the packet. The Supervisor Engine also informs (over the Results Bus (RBus)) all the nondestination ports on the switch to flush the packet from their buffers.
- The RBus also conveys to the destination port or ports (in the case of multicast traffic) the outgoing Ethernet frame MAC address rewrite information (source MAC and next hop MAC addresses) and the relevant QoS instructions to be used to queue the packet correctly on the exit port.
- Upon receiving the packet, the PINNACLE ASIC on the destination port places the packet in the correct TX queue. The ASIC then uses its strict priority and WRR schedulers to transmit the framed packet out of its buffer on its way to the destination.

8

ARCHITECTURES WITH SHARED-MEMORY-BASED SWITCH FABRICS: CASE STUDY—CISCO CATALYST 3550 SERIES SWITCHES

8.1 INTRODUCTION

The Cisco Catalyst 3550 Series are fixed configuration, stackable switches that employ a distributed shared memory switch fabric architecture. The architecture of the Catalyst 3550 is based on the older Catalyst 3500 XL Layer 2 (only) switch [CISC3500XL99]. The Catalyst 3500XL switch is one of the switches that belongs to the Catalyst "XL" family of switches that includes the Catalyst 2900XL and Catalyst 2900XL LRE. The XL family of switches are strictly Layer 2 switches, with no Layer 3 capabilities beyond the simple functions provided by the management interface (Telnet, SNMP, etc.).

The Catalyst 3550 switches are enterprise-class switches that support Layer 2 and 3 forwarding as well as quality of service (QoS) and security features required in many of today's networks. The Catalyst 3550 Series support a range of 100 Mb/s Ethernet and Gigabit Ethernet interfaces that allow them to serve as access layer switches for medium enterprise wiring closets and as backbone switches for medium-sized networks.

In the Catalyst 3550 [CISC3550DS05, CISC3550PRS03, CISCRST2011], all Layer 2 and 3 forwarding decisions are performed in network interfaces modules ASICs (referred to as satellite ASICs). The Layer 2 and 3 forwarding decisions in some cases involve processing Layer 4 parameters of the arriving packets. The network satellite ASICs manage either a group of 100 Mb/s Ethernet ports or a

Switch/Router Architectures: Shared-Bus and Shared-Memory Based Systems, First Edition. James Aweya.
© 2018 The Institute of Electrical and Electronics Engineers, Inc. Published 2018 by John Wiley & Sons, Inc.

single Gigabit Ethernet (GbE) port. A central CPU in the Catalyst 3550 is responsible for running the Layer 2 and 3 protocols, routing table management, and overall system control and management.

The Catalyst 3550-12 T/12G has a 24 Gb/s switch fabric capacity and supports 17 million packets per second (Mpps) throughput, Catalyst 3550-24 has 8.8 Gb/s capacity with 6.6 Mpps throughput, and Catalyst 3550-48 has a 13.6 Gb/s capacity with 10.1 Mpps throughput. The Catalyst 3550-24 supports a 2 MB shared memory shared by all switch ports, 64 MB RAM, 16 MB Flash memory, storage of 8000 MAC addresses, 16,000 unicast routes, 2000 multicast routes, and maximum transmission unit (MTU) of 1546 bytes for MPLS forwarding.

Based on architecture categories described in Chapter 3, the architecture discussed here falls under "Architectures with Shared-Memory-Based Switch Fabrics and Distributed Forwarding Engines" (see Figure 8.1).

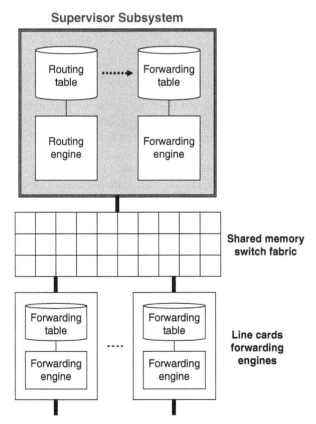

FIGURE 8.1 Architecture with shared-memory-based switch fabric and distributed forwarding engines.

8.2 MAIN ARCHITECTURAL FEATURES OF THE CATALYST 3550 SERIES

In the Catalyst 3550, a switching and packet forwarding subsystem supports a shared memory switch fabric ASIC that manages transactions between a centralized 4 MB shared memory buffer and a number of network interface modules (referred to as network satellites in the Catalyst 3550) interconnected in a radial design (Figure 8.2). The Catalyst 3550 supports 10 Gb/s of raw bandwidth capacity between the shared memory switch fabric ASIC and the shared (data) memory buffer. This yields a data forwarding rate of 5 Gb/s in one direction.

In the Catalyst 3550, the network satellites provide the interfaces to the external network. Each satellite performs the address lookups and forwarding for incoming packets using its own address table. The Catalyst 3550 supports two network satellite types (octal 10/100 Ethernet satellite and a single-port Gigabit Ethernet satellite) and each satellite handles all addressing operations for incoming traffic. The network satellites communicate with each other by sending notification messages over the notify ring, which is more efficient than traditional bus architectures, potentially delivering up to 10 million frame notifications per second.

Depending on the traffic load, a network satellite is allowed to use (dynamically) all or some of this shared memory switching bandwidth. All incoming packets pass through the shared memory switch fabric ASC and are stored in the shared memory data buffer. A shared memory architecture eliminates the "head-of-line" blocking problems normally associated with pure input-buffered architectures.

The Catalyst 3550 supports radial (store/receive) channels that connect the shared memory switch fabric ASIC and the network satellites. Each channel

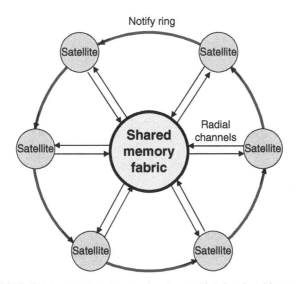

FIGURE 8.2   Catalyst 3550 switch/router high-level architecture.

provides 200 Mb/s bandwidth in each direction resulting in a total full-duplex channel capacity of 400 Mb/s between each satellite and the switch fabric ASIC.

8.3 SYSTEM ARCHITECTURE

Figure 8.2 presents a high-level architecture of the Catalyst 3550 series of switch/ routers. This architecture was developed to strike a good balance between obtaining maximum packet forwarding performance in hardware and software design flexibility. A more detailed presentation of the architecture is given in Figure 8.3.

8.3.1 Packet Forwarding Subsystem

At the core of the Catalyst 3550 switch/router architecture is the switching and packet forwarding subsystem (see Figures 8.2 and 8.3). This subsystem consists of the shared memory switch fabric ASIC, network satellites (module port and octal Ethernet satellites) that act as network interface modules, shared data memory buffer, and notify ring.

8.3.1.1 Switching and Forwarding Engines   The switching and forwarding engines implemented in the network satellites handle the primary packet forwarding functions, including receiving and transmitting user data traffic. The switching and forwarding engines provide low-latency, high-performance Layer 2 ad 3 forwarding and allow all destination address lookups to be performed entirely in (distributed) network satellites. The initial implementations of the Catalyst 3550 switch/ router architecture supported 10/100 Mb/s and Gigabit Ethernet ports.

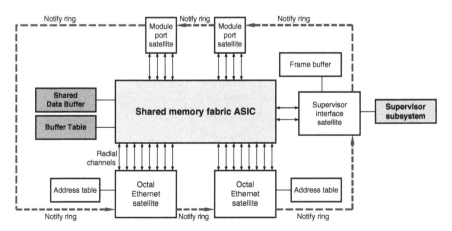

FIGURE 8.3   Catalyst 3550 switch/router architecture details.

8.3.1.2 Shared Memory Switch Fabric ASIC The shared memory switching fabric ASIC (Figure 8.3) is responsible for managing its associated shared data buffer and buffer table. The buffer table maintains addressing information used by the shared data buffer. The 10 Gb/s link interconnecting the shared memory switch fabric ASIC and the shared data buffer provides a 5 Gb/s forwarding rate.

The radial channels (Figures 8.2 and 8.3) that connect network satellites to the shared memory switch fabric ASIC distribute the total available system bandwidth among the network satellites. The radial channels are designed to minimize the number of pins needed per data to maximize system reliability and lower cost.

8.3.1.3 Shared Data Buffer The shared data buffer is a key component of the shared-memory-based Catalyst 3550 architecture. The shared data were based on a 4 MB DRAM in the initial deployment of the Catalyst 3550. A shared data buffer architecture allows the Catalyst 3550 to optimize buffer utilization (especially under varying network traffic loads) through dynamic buffer allocation to all the system ports. The shared data buffer also allows the system to avoid duplicating multicast or broadcast packets to the destination ports.

The shared data buffer bandwidth also provides an efficient use of memory bandwidth and storage capacity. The shared buffer allows designers to reduce the total amount of memory required in the switch/router, while providing high nonblocking performance. All incoming packets are temporarily stored in a common "memory pool" until the destination ports are ready to read and transmit the packets. Being a shared resource, heavily loaded destination ports can consume as much memory as they need, while lightly loaded ports do not have to hog unused memory space.

The shared memory also allows larger bursts of traffic from a port than corresponding port-buffered architectures. With a good combination of adequate buffering with dynamic allocation, this architecture effectively eliminates or reduces significantly packet loss during traffic overload due to limited buffer capacity. Similarly, with adequate buffering and dynamic buffer allocation, the system avoids head-of-line blocking problems normally associated with input-buffered architectures without per output port buffers (also known as virtual output queues (VoQs)).

Unlike input-buffered architectures with VoQs (that must store multiple copies of a multicast or broadcast packet (one in each destination VoQ)), shared buffer architectures increase the overall system performance by eliminating the replication of multicast and broadcast packets. The shared memory switch fabric ASIC maintains logical queues in the buffer table that are dynamically linked to transmit queues for each destination port, with multiple references to the same buffer location for a multicast or broadcast packet. A multicast packet (which is destined for multiple destination system ports and network addresses) is stored in the same shared memory location until all destination ports have read and forwarded their copies of the packet.

8.3.1.4 Network Satellites   The network satellites provide connectivity to the external network and also manage media interfaces to the network. The satellites transfer and receive packets from the shared data buffer and perform Layer 2 and 3 destination address lookups in their local forwarding tables. The network satellites are also responsible for 10/100 Mb/s Ethernet (and other media) MAC (Media Access Control) protocol functions, determining source/destination Layer 2 (MAC) addresses of incoming packets, updating and synchronizing the local Layer 2 (address) tables, and supporting up to 200 Mb/s full-duplex data transfer from each network port to the shared memory switch fabric ASIC.

To ensure data integrity, the local address table in any network satellite is synchronized with the tables in other network satellites via the notify ring. When a packet arrives at a network satellite, it converts the incoming packet into fixed-length cells and transfers them to the shared memory switch fabric ASIC for storage in the shared data buffer. At the same time, the source network satellite performs a destination address lookup in its local forwarding table and notifies the destination ports via the notify ring interconnecting the network satellites.

The destination port receives the notification, and then reads and reconverts the cells belonging to the outgoing packet into a complete packet before forwarding it out of the port. The types and number of network satellites employed in the Catalyst 3550 vary depending on its implementation. Each 100 Mb/s Ethernet network satellite can support up to eight independent 10/100BASE-T Ethernet ports (referred to as octal Ethernet module in Figure 8.3), while each Gigabit Ethernet network satellite (i.e., the module port satellite in Figure 8.3) supports only one 1000BASE-X Ethernet port.

8.3.1.5 Notify Ring   The notify ring carries notifications between network satellites and also management information for the synchronization of the address tables, confirmation of packet arrival at a satellite, notification of packet retrieval by a satellite, and other operational related activities among the satellites. The notify ring provides an effective way to off-load communications between the network satellites (i.e., "housekeeping" tasks) to a dedicated "out-of-band" channel associated with the existing switch fabric. This approach offsets the packet forwarding performance degradation that may have occurred if the switch/router had integrated all these functions into a system common channel.

Each packet notification message contains a queue map that is read by each network satellite and after which it is forwarded on the ring to the next module. This notification message may carry information about packet type, its queuing priority, and so on. With this, the amount of queue numbers in a notification message can exceed the number of ports in a system. When a notification message carries information relevant to a particular network satellite, it modifies the message in response and then forwards it on the notify ring.

The notify ring is designed to be an 800 Mb/s, 8 bit unidirectional communication ring interconnecting all the network satellites. The notify ring has a notification message size of 10 bytes per packet, thus resulting in the Catalyst 3550 supporting up to 10 million packet notifications per second.

8.3.1.6 Radial Channels The shared memory switch fabric ASIC communicates with other components in the system and the network satellites through the radial channels. The number of radial channels varies according to the Catalyst 3550 switch/router design, but each radial channel consists of four unidirectional signal pathways (subchannels). Two signal pathways are used for incoming data storage and two signal pathways for outgoing data retrieval. Each signal pathway set also carries all in-band signaling. Each radial channel can support up to 200 Mb/s of data in each direction simultaneously. Excluding control and overhead traffic, a typical radial channel has approximately 160 Mb/s of full-duplex payload capacity.

8.3.1.7 Format Conversions Data are stored in the shared data buffer, read, and moved across the radial channels in fixed-length cells. The network satellites are responsible for carrying out the conversion of incoming packets to be transported and stored in the shared data buffer. The fixed-length cells make transfer and storage more predictable and enable the switch/router manage the shared data buffer more efficiently. A header attached to data is read and interpreted by the shared memory switch fabric ASIC during storage (in the shared data buffer) and by the network satellite during data retrieval.

The data headers identify the origin of a frame (packet) and its boundaries, the number of expected reads/retrievals (from memory), and other information needed for handling the packet. When storing a cell carrying a segment of a packet in the shared data buffer, the shared memory switch fabric ASIC reads the data header to create a temporary address entry in the buffer table.

8.3.1.8 Destination Address Lookup When a network satellite receives a packet, it stores it (via the shared memory switch fabric ASIC) in the shared data buffer. The network satellite performs a lookup in its local address table to determine the packet's destinations. The source network satellite then notifies the destination satellites by sending notifications over the notify ring. The packet is segmented into cells, each one with a header when stored in the shared data buffer. The number of destinations for the packet is contained in the cell header and includes a retrieval count (that indicates which destinations have copied the cell so far).

When a destination network satellite receives a notification message from the source satellite, it reads and appends the information sent to its local notify queue. The destination satellite then signals the source satellite that it is ready to retrieve and forward the cells that make up the packet. Once retrieved, the destination satellite reassembles the cells into the full packet and forwards it through the appropriate local ports. If a destination network satellite is not able to accept more packets, the source network satellite notifies the shared memory switch fabric

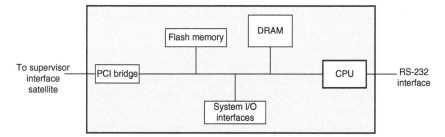

FIGURE 8.4   Supervisor Subsystem–CPU interface satellite.

ASIC, which adjusts (e.g., delete) the entry in the buffer table for each packet that cannot be sent.

8.3.2   Supervisor Subsystem

Figure 8.4 shows the architecture of the supervisor subsystem of the Catalyst 3550 switch/router. The supervisor subsystem connects to the shared memory fabric ASIC of the Catalyst 3550 via a supervisor interface satellite as illustrated in Figure 8.3. This subsystem contains a control CPU, Flash memory, DRAM, system input/output (I/O) interfaces, PCI bridge, and serial (RS-232) interface ports (for system management). The supervisor subsystem supports higher level protocols and applications used to control, monitor, and manage the overall Catalyst 3550 switch/router.

The various components of the supervisor subsystem are described as follows.

8.3.2.1 *Control CPU*   The control CPU is a 32 bit PowerPC RISC processor that provides Layer 3 functions such as routing protocol processing, Layer 2 functions (e.g., Rapid Spanning Tree Protocol (RSTP), IEEE 802.1AB Link Layer Discovery Protocol (LLDP), and VLAN Trunking Protocol (VTP)), Layer 3 routing table construction and maintenance, Layer 2 address table maintenance, connection management, and network management functions.

When the switch/router is powered on, the control CPU automatically initiates a self-diagnosis of the systems and other system control tasks. The Catalyst 3550 supports management features such as SNMP, Telnet, Cisco Visual Switch Manager (CVSM), and command-line interface (CLI). The Catalyst 3550 supports four groups of RMON and security features.

8.3.2.2 *Supervisor Interface Satellite*   The supervisor interface satellite provides connectivity between the supervisor subsystem and the shared memory fabric ASIC of the switch/router. This interface provides a channel between the switch fabric resources and the control CPU and its support components (Flash memory, system I/O interfaces, and serial interface ports). The supervisor interface satellite formats address tables used by network satellites (module port and octal Ethernet satellites).

8.3.2.3 Flash Memory   This is a nonvolatile Flash memory of 4 MB in size used to store the Catalyst 3550 Cisco IOS software image, current switch/router (system) configuration information, and a built-in CVSM software. A true file system with directory structures is supported in the Flash memory that allows easy software upgrades. The Flash memory maintains stored information across power cycles, thus facilitating maximum system reliability.

8.3.2.4 System I/O Interface   The system I/O interfaces are used to provide control and status for various system-level functions such as system status, LED control, an RS-232 (also known as EIA/TIA-232) serial interface (that allows access from a system console device for management purposes), and an external redundant power supply interface.

8.4 PACKET FORWARDING

In centralized forwarding, a single central forwarding engine is used that performs all forwarding operations (Layer 2, Layer 3, QoS, ACLs, etc.) for the system. Here, the system performance is determined by the performance of the central forwarding engine. In a distributed forwarding architecture like the Catalyst 3550, the switching and forwarding decisions are made at module or port level with local forwarding engines and forwarding tables.

These distributed forwarding tables are synchronized across all the distributed forwarding engines to allow consistent forwarding decisions in the system. The overall system performance is equal to the aggregate performance of all forwarding engines in the system. Distributed forwarding allows switches, switch/routers, and routers to achieve very high packet forwarding performance.

In flow-based forwarding (also known as demand-based forwarding), forwarding is based on traffic flows where the first packet is forwarded in software by the route processor. Subsequent packets of the same flow are then forwarded in hardware by forwarding engine ASICs using the flow cache created. A flow can consist of the source address, source/destination addresses, or full Layer 3 and Layer 4 information. The scalability of flow-based forwarding is dependent on the control plane performance. Issues such as the following have to be addressed when implementing flow-based forwarding:

- How fast the route processor can process new flows and set them up in the forwarding engine hardware?
- How network topology changes (including route flaps, etc.) are handled and managed in the flow cache?
- Given that the route processor is responsible for control plane functions, the other tasks it is responsible for (other than routing protocols, ARP, spanning tree, etc.) can affect the processing power devoted to flow processing.

- The stability of the critical routing protocols processes, while flows are being established in the route processor, has to be ensured.

The Catalyst 3550 uses topology-based forwarding tables where the Layer 3 forwarding information is derived from the routing table maintained from the routing protocols and the node adjacencies are derived from the Address Resolution Protocol (ARP) table. In this architecture, the Layer 3 forwarding tables (including the adjacency information) are generated and built from the system control plane.

These tables are installed in the ASIC hardware in the network satellites of the Catalyst 3550. The lookup in the forwarding table is based on destination longest-match prefix search. A forwarding table hit returns an adjacency (next hop IP node), outgoing port, and adjacency rewrite information (next hop MAC address).

In a distributed forwarding system, the scalability of the system is dependent on the forwarding engines' performance and not on flow-based hardware forwarding of first packet in each flow, no matter how many flows exist in the system (whether there are one or one million new flows). In this system, the hardware forwarding tables are identical to software tables maintained by the central route or control CPU.

The hardware forwarding tables are updated by the routing protocol software in the control CPU as network topology changes occur. The control plane is decoupled from normal user traffic forwarding and dedicated to protocol processing (routing protocols, ARP, spanning tree, etc.).

8.4.1 Catalyst 3550 Packet Flow

This section describes the packets forwarding process in the Catalyst 3550. The processing steps are described in Figures 8.5 and 8.6:

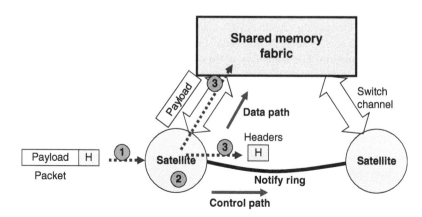

FIGURE 8.5   Packet flow – ingress.

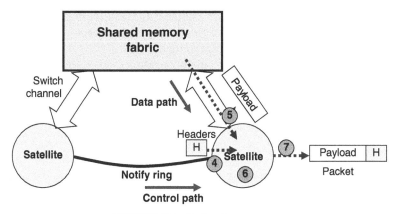

FIGURE 8.6   Packet flow – egress.

1. A packet arrives from the external network to a port on a network satellite.
2. The ingress network satellite ASIC makes the relevant Layer 2 or Layer 3 forwarding decisions (plus policing, marking, etc.).
3. The ingress network satellite parses the packet header from the payload and sends the following:
   a. Header information on the notify ring to the egress ports (this is the control path).
   b. Packet payload to the shared memory switch fabric ASIC for temporary storage in the shared data buffer (this is the data path).
4. The egress network satellite receives the control information on the notify ring and recognizes that it is one of the destination ports.
5. The egress network satellite then retrieves the packet from the shared data buffer for all its local destination ports.
6. The egress network satellite performs packet rewrite (on the relevant Layer 2 and 3 header fields), output ACL filtering and policing, and local multicast expansion.
7. The egress network satellite transmits the packet out the local egress port(s).

The Catalyst 3550 uses a TCAM (ternary content-addressable memory) for storing the forwarding information required for forwarding traffic. The available TCAM space is shared among all forwarding entries in the system. Sharing of these forwarding entries is based on predefined templates, where the templates "carve" out the TCAM space to suit the network environment, for example, routing and VLAN.

8.4.2   Catalyst 3550 QoS and Security ACL Support

The Catalyst 3550 and 3750 series switches supports router-based access control lists (RACLs), VLAN-based ACL (VACLs), and port-based ACL (PACL). The

Catalyst 3550 supports 256 security ACLs on the 10/100 Ethernet satellites with 1 K security ACEs (access control entries). The security ACLs programmed in TCAM are used for hardware enforcement of security policies.

RACLs can be applied on switch virtual interfaces (SVIs) (see SVIs below), which are routed (Layer 3) interfaces to VLANs on the Catalyst 3550, on physical routed interfaces, and on routed (Layer 3) EtherChannel interfaces. RACLs are applied in specific directions on interfaces (inbound or outbound) where the user can apply one IP ACL in each direction.

VLAN maps (or VACLs) can be applied on the Catalyst 3550 to all packets that are Layer 3 forwarded (routed) into or out of a VLAN or are Layer 2 forwarded within a VLAN. VACLs are mostly used for security packet filtering. PACL can also be applied to Layer 2 interfaces on the Catalyst 3550. PACLs are supported on physical ports/interfaces only and not on EtherChannel interfaces.

The ACLs supported on the Catalyst 3550 are summarized below:

- **Router ACL (RACL)**
  - Applied to routed ports and SVI.
  - Standard and Extended IP ACLs.
  - Can be applied to data plane or control plane traffic on all ports.
  - Filter on Source/Destination MAC address, Source/Destination IP address, and TCP/UDP port numbers.
- **Port ACL (PACL)**
  - Applied to specific switch port.
  - Filter on Source/Destination MAC address, Source/Destination IP address, and TCP/UDP port numbers.
- **VLAN ACL (VACL)**
  - Applied to all packets either bridged or routed within a VLAN, including all non-IP traffic.
  - Filter on Source/Destination MAC address, Source/Destination IP address, and TCP/UDP port numbers.
- **ACL Hierarchy:** On the ingress interface, the VLAN ACL gets applied first. On the egress interface, it is applied last.
- **Time-Based ACLs**: These are security ACLs set for specific periods of the day.

The Catalyst 3550 supports the following QoS features [CISCRST2011, CISCUQoS3550, FROOMRIC03]. Chapters 11 to 13 describe these QoS mechanisms in detail:

- **Scheduling:**
  - Egress scheduling
  - Strict priority queuing

- Egress weighted round-robin (WRR) with weighted random detection (WRED)
- **Traffic Classification and Marking**
  - Based on default port IEEE 802.1Q (sometimes referred to as 802.1p) class of service (CoS) or Layer 2/Layer 3/Layer 4 ACL policies.
  - 512 QoS ACEs supported on all 10/100 Ethernet configurations.
  - IEEE 802.1Q (CoS), Cisco inter-switch link (ISL), Differentiated Services Code Point (DSCP), or IP Precedence marking.
- **Rate Policing**
  - Policer support:
    o 128 ingress policers per Gigabit Ethernet port.
    o Eight ingress policers per 100 Mb/s Ethernet port.
    o Eight egress policers per 100 Mb/s and Gigabit Ethernet ports.
  - Support of per interface and shared aggregate policers.

The header of an ISL frame (which are Layer 2 frames) contains a 1 byte user field with the three least significant bits used to carry an IEEE 802.1p CoS value. Interfaces configured as ISL trunks format and transport all traffic in ISL frames. The header of an IEEE 802.1Q frame (also a Layer 2 frame) contains a 2 byte Tag Control Information (TCI) field with the three most significant bits (called the Priority Code Point (PCP) bits) used to carry the CoS value. Except for traffic in the native VLAN, interfaces configured as IEEE 802.1Q trunks format and transport all traffic in IEEE 802.1Q frames. The ISL and IEEE 802.1Q CoS field take values from 0 for low priority to 7 for high priority.

Cisco ISL was developed as an encapsulation protocol to allow multiple VLANs to be supported over a single link (trunk). With ISL, Ethernet frames are encapsulated with the VLAN information and a 26 byte header and a new 4 byte CRC (at the end of the ISL packet) are added. On the ISL trunk port, all packets received and transmitted are encapsulated with an ISL header. Nontagged or native frames received on an ISL trunk port are discarded.

A native VLAN on an IEEE 802.1Q trunk is the only untagged VLAN (only VLAN that is not tagged in the trunk). Frames transmitted on a switch port on the native VLAN are not tagged. Generally, if untagged frames are received on a switch on a IEEE 802.1Q trunk port, they are assumed to be from a VLAN that is designated as the native VLAN. The native VLAN is not necessarily the same as the management VLAN; they are generally kept separate for better security.

Layer 3 (IP) packets can be marked with CoS values using either IP Precedence or a DSCP marking. The Catalyst 3550 supports the use of either CoS settings because DSCP values are backward-compatible with IP Precedence values. IP Precedence values range from 0 to 7, while those for DSCP range from 0 to 63.

The Catalyst 3550 supports features to classify, reclassify, police, and mark arriving packets before they are stored in the shared data buffer via the shared memory switch fabric ASIC. Packet classification mechanisms allow the Catalyst 3550 to differentiate between the different traffic flows and enforce QoS and security policies based on Layer 2 and Layer 3 packet fields.

To implement QoS and security policies at the ingress, the Catalyst 3550 identifies traffic flows and then classifies/reclassifies these flows using the DSCP or IEEE 802.1Q CoS fields. Classification and reclassification can be based on criteria such as the source/destination IP address, source/destination MAC address, and TCP or UDP ports. The Catalyst 3550 will also carry out policing and marking of the incoming packets. In addition to data plane ACLs, the Catalyst 3550 supports control plane ACLs on all ports to ensure that packets destined to the route processor (i.e., supervisor subsystem) are properly policed and marked to maintain proper functioning of the routing protocol processes.

After the packets are classified, policed, and marked, they are then assigned to the appropriate priority queue before they are transmitted out the switch. The Catalyst 3550 supports four egress priority queues per port (one of which is a strict priority queue), which allows the switch/router to assign priorities for the various traffic types transiting the switch/router. At egress, the Catalyst 3550 performs scheduling of the priority queues and also implements congestion control. The Catalyst 3550 supports strict priority queuing and WRR scheduling (on the remaining three queues).

With the WRR scheduling, the Catalyst 3550 ensures that the three lower priority queued packets are not starved of output link bandwidth and are serviced proportional to the weights assigned to them. Strict priority queuing allows the Catalyst 3550 to ensure that the (single) highest-priority queued packets will always get serviced first, before the other three queues that are serviced using WRR scheduling. In addition to these scheduling mechanisms, the Gigabit Ethernet ports on the Catalyst 3550 support congestion control via WRED. WRED allows the Catalyst 3550 to avoid congestion by allowing the network manager to set thresholds on the three lower priority queues, at which packets are dropped before congestion occurs.

The Cisco Catalyst 3550 supports a Cisco Committed Information Rate (CIR) functionality that is used to perform rate limiting of traffic. With CIR, the Catalyst 3550 can guarantee bandwidth in increments of 8 kbps. Bandwidth guarantees can be configured based on criteria such as source MAC address, destination MAC address, source IP address, destination IP address, and TCP/UDP port numbers. Bandwidth guarantees are an essential component of service-level agreements (SLAs) and in networks the network manager needs to control the bandwidth given to certain users.

Each 10/100 port on the Catalyst 3550 supports eight individual ingress policers (or eight aggregate ingress policers) and eight aggregate egress policers. Each Gigabit Ethernet port on the Catalyst 3550 supports 128 individual ingress policers (or 128 aggregate ingress policers) and 8 aggregate egress policers. This allows the network manager to implement policies with very granular control of the network bandwidth.

8.5 CATALYST 3550 SOFTWARE FEATURES

The Cisco Catalyst 3550 Series switches support advanced features, such as advanced QoS management and control, rate-limiting ACLs, multicast traffic management, and advanced IP unicast and multicast routing protocols. Supported in the Catalyst 3550 is the Cisco Cluster Management Suite (CMS) Software, which allows network managers to configure and troubleshoot multiple Catalyst switches (switch cluster) using a standard Web browser.

The routing protocols include Routing Information Protocol v1/2 (RIPv1/v2), Open Shortest Path First (OSPF), Interior Gateway Routing Protocol (IGRP), Enhanced Interior Gateway Routing Protocol (EIGRP), Border Gateway Protocol version 4 (BGPv4), Protocol Independent Multicast (PIM), Internet Group Management Protocol (IGMP), and Hot Standby Router Protocol (HSRP). The Catalyst 3550 also supports IGMP snooping in hardware, which makes the switch/router effective for intensive multicast traffic environments. Additionally, the Catalyst 3550 supports equal cost routing (ECR) with load balancing on routed uplinks to allow for better bandwidth utilization. The Catalyst 3550 supports mechanism for performing load balancing on the routed uplinks.

Each individual port on the Catalyst 3550 can be configured as a Layer 2 interface or a routed (Layer 3) interface. A Layer 3 interface is a physical port that can route/forward Layer 3 (IP) traffic to another Layer 3 device. A routed (Layer 3) interface does not support and participate in Layer 2 protocols, such as the Rapid Spanning Tree Protocol (RSTP). When a port is configured to act as a routed interface, its protocol functions are no different than configuring a 100 Mb/s or Gigabit Ethernet port on a router. An IP address can be assigned to this interface, as well as ACLs- and QoS-related configurations can be applied. The network manager can assign an IP address to the Layer 3 interface, enable routing, and assign routing protocol characteristics to this Layer 3 interface.

VLAN interfaces (or Switched Virtual Interfaces (SVIs)) can also be configured on the Catalyst 3550. An SVI is a virtual (logical) Layer 3 interface that connects the routing (Layer 3) engine on a device to a VLAN configured on the same device. Only one SVI can be associated with a VLAN on the device. An SVI is configured for a VLAN only when there is the need to route between VLANs.

An SVI can also be used to connect the device to another external IP device through a virtual routing and forwarding (VRF) instance that is not configured as a management VRF. A device can route across its SVIs to provide Layer 3 inter-VLAN communications/routing. This requires configuring an SVI for each VLAN that traffic is to be routed to and assigning an IP address to the SVI.

To summarize, the interfaces supported on the catalysts 3550 are as follows:

- **Switch Ports:** These are Layer 2-only interfaces on the switch with one interface per physical port:
  - **Access Ports**: Traffic received and transmitted over these ports must be in native format (i.e., VLAN-tagged traffic is dropped).

- **Trunk Ports:** These ports carry traffic from multiple VLANs:
- **ISL-Trunks:** Packets over these trunks must be encapsulated with an ISL header.
- **IEEE 802.1Q-Trunks**: VLAN-tagged packets are trunked over these trunks but untagged packets are sent to native VLAN (or user-defined default VLAN).
- **Layer 3 (Routed) Ports:** These ports are configured to behave like traditional router ports.
- **VLAN Interface (or Switch Virtual Interface (SVI)):** This interface provides a connection between a Layer 3 routing process and an attached switched VLAN (a Layer 2 bridged access to a VLAN).

The Catalyst 3550 supports a number of security features to prevent unauthorized access to it. Access can be via issuing passwords on a console and VTY lines, username/password pairs stored locally on the Catalyst 3550 for individual access, and username/password pairs stored on a centrally located TACACS+ or RADIUS server. A virtual teletype (VTY) is a CLI implemented in a device that facilitates accessing it via Telnet, for example.

Privilege levels can also be configured for passwords where a user can be granted access at a predefined privilege level when the user enters the correct password. To support the ability to give different levels of configuration capabilities to different network managers, the Catalyst 3500 has 15 levels of authorization on the switch/router console and 2 levels on a Web-based management interface.

The Catalyst 3550 support a wide range of security features (e.g., Secure Shell (SSH), Simple Network Management Protocol version 3 (SNMPv3), Kerberos) that protect administrative and network management traffic from tampering or eavesdropping. The switch/router supports features and protocols that can encrypt administrative and network management information to allow secure communications with users and other devices.

- **Secure Shell (SSH):** SSH encrypts administration traffic during Telnet sessions while the network administrator configures or troubleshoots the switch.
- **SNMPv3 (with Crypto Support):** Provides network security by encrypting network administrator traffic during SNMP sessions to configure and troubleshoot the switch.
- **Kerberos:** Provides strong authentication for users and network services using a trusted third party to perform secure verification.

For secure, remote connection to the Catalyst 3550, SSH can be used. User authentication methods can be via TACACS+, RADIUS, and local username authentication. A RADIUS client runs on the Catalyst 3550 and sends authentication requests to a central RADIUS server, which contains all information on user authentication and network service access. TACACS+ provides centralized validation of users seeking to gain access to the Catalyst 3550. The TACACS+ services are maintained in a database on a TACACS+ server running on a workstation. The TACACS+ server has to be configured before the configuring TACACS+ features on the Catalyst 3550. TACACS+ is modular and provides authentication, authorization, and accounting services separately. With TACACS+, the TACACS+ server (implemented in a single access control device) is able to provide each authentication, authorization, and accounting service, separately and independently.

The Catalyst 3550 supports IEEE 802.1X that defines a client-server access control and authentication protocol that restricts unauthorized clients from accessing a network through the Catalyst 3550. The authentication server performs the authentication of each client connected to a Catalyst 3550 port before permitting access to any services offered by the switch or the network. A user can be authenticated using IEEE 802.1X based on a username and password (or other credentials supplied by the user) through a RADIUS server.

8.6 CATALYST 3550 EXTENDED FEATURES

The Catalyst 3550 supports a number of extended and advanced features beyond Layer 2 and 3 packet forwarding, some of which are discussed here.

8.6.1 EtherChannel and Link Aggregation

The Catalyst 3550 also supports 100 Mb/s Ethernet and Gigabit EtherChannel, which is a port link aggregation technology developed by Cisco. EtherChannel (similar to IEEE 802.3ad Link Aggregation) allows several physical Ethernet links to be grouped to create one logical Ethernet link. This provides high-speed links with fault-tolerance between switches, switch/router, routers, and servers in a network.

EtherChannel technology allows a network manager to aggregate multiple 100 Mb/s Ethernet or Gigabit Ethernet links to create a higher bandwidth connection (with scalable bandwidth and higher availability) between switches, servers, switch/routers, and routers than a single 100 Mb/s or Gigabit Ethernet links can provide. In the Catalyst 3550 with EtherChannel technology, all incoming packets (to the network satellites) are stored in the shared data buffer in the order in which they arrive, but are properly resequenced by the network satellites when forwarded. In the Catalyst 3550, EtherChannel provides full-duplex bandwidth up to 800 Mb/s (100 Mb/s EtherChannel) or 8 Gb/s (Gigabit EtherChannel) between the Catalyst 3550 and another device.

8.6.2 Port Security

The port security feature can be used to restrict access to a port by identifying and limiting MAC addresses (end stations) that are authorized/unauthorized to access the port. When "secure" or "trusted" MAC addresses are assigned to a port, the port only forwards packets with source addresses from that group of specified MAC addresses. If the number of trusted MAC addresses is limited to one and only a single trusted MAC address is assigned to a port, then the owner of that address is ensured the full bandwidth of the port.

If a port is configured as a trusted port and the maximum number of trusted MAC addresses is assigned, a security violation occurs when other MAC address (different from any of the specified trusted MAC addresses) attempt to access the port. After the maximum number of trusted MAC addresses have been set on a port, the trusted addresses are added to the address table manually (statically) by the network manager or the port dynamically configures trusted MAC addresses with the MAC addresses of connected stations. However, if the port shuts down, all dynamically learned trusted MAC addresses are lost/deleted. After the maximum number of trusted MAC addresses has been configured (manually or dynamically), they are stored in an address table.

A port with port security configured with "sticky" trusted MAC addresses provides the same benefits as port security configured manually (i.e., with static MAC addresses), but with the exception that sticky MAC addresses can be dynamically learned. These sticky MAC addresses are stored in the address table, and added to the switch's running configuration.

The sticky trusted MAC addresses (even if added to the running configuration) do not automatically become part of the switch's start-up configuration file (each time the switch restarts or reboots). If the sticky trusted MAC addresses are saved in the start-up configuration file, then when the switch restarts/reboots, the port does not need to relearn these addresses. The switch retains these dynamically learned MAC addresses during a link-down condition. If the start-up configuration is not saved, they are lost when the system restarts/reboots.

As noted above, a security violation occurs if a station with a MAC address not in the address table attempts to access the port. A port can be configured to take the following actions if a violation occurs. A port security violation can cause an SNMP notification to be generated and sent to a management station or it can cause the port to shut down immediately.

To manage and control unauthorized access to the Catalyst 3550 switch/router (port security), a network manager can configure up to 132 "trusted" MAC addresses per port. When port security is configured on the switch/router, the network satellite that handles a port applies the filtering policies as part of the normal address learning and filtering process. Switch/router ports know about trusted MAC addresses either through manual configuration by a network manager or automatically through the connected end stations.

When automatic configuration is used, the network manager waits until a port goes through "learning" the MAC addresses of the connected devices, and then after a period of time "freezes" the trusted address table. Only packets from these "trusted" MAC addresses (maintained in the address table) are granted access through the port. If a port security violation occurs, the port can block access or ignore the violation and send an SNMP trap to a management station. Port security can be configured using the Web-based CVSM interface on the Catalyst 3550.

8.6.3 Switch Clustering

Switch clustering can be used to simplify the management of multiple switches in a network, regardless of device/platform family and their physical geographic proximity. Through the use of standby cluster command switches, a network manager can also use clustering to provide switching redundancy in the network. A switch cluster can consist of up to 16 cluster-capable switches that are managed as a single logical switching entity. The cluster switches can support a switch clustering technology that allows configuring and troubleshooting them as a group through a single IP address. The external network communicates with the switch cluster through the single IP address of the cluster command switch.

In the switch cluster (a cluster cannot exceed 16 switches), one switch is designated the cluster command switch, and one or more other switches (up to 15) can be designated as cluster member switches. The role of the cluster command switch is to serve as the single point of access for configuring, managing, and monitoring the cluster member switches. However, cluster member switches can belong to only one switch cluster at a time since they cannot be configured, managed, and monitored by more than one cluster command switch at the same time.

More than one cluster member switch can be designated as standby cluster command switch to implement command switch redundancy if the active cluster command switch fails. This is to avoid loss of communication with cluster member switches when the active command switch fails. A cluster standby group can also be configured for a switch cluster that consists of a group of standby cluster command switches.

The Catalyst 3550 switch/router supports the Cisco switch clustering technology that enable up to 16 switches to be managed (logically as one unit) through a single IP address (independent of the media interconnecting them or their geographic proximity).

8.6.4 Channel Multiplexing and Frame Stitching

The Catalyst 3550 supports other advanced packet switching and forwarding features such as channel multiplexing and frame (packet) "stitching." Channel multiplexing is the Catalyst 3550's ability to support multiple "threads" (up to 256 threads) per radial channel. Each cell of an arriving packet (frame) contains a thread identifier that is used to multiplex and demultiplex data transferred over the radial channel. This capability may be applied to network interface modules that support multiple 100BASE-T ports associated with a single radial channel.

The Catalyst 3550 uses frame stitching to modify packets "on the fly" without degrading overall packet forwarding performance. With this, a network satellite would be able to read part of a packet (usually the cell containing the packet header), process and modify it, and write back the modified header in the shared data buffer. The satellite would then edit the contents of the buffer table to "stitch" the new "header" cell into the old packet, effectively overwriting the original first "header" cell.

This allows packets with modified headers to be created without the need to retrieve, modify, and write back entire packets. After frame stitching occurs, the source network satellite transmits a frame-notify message via the notify ring to the destination satellite to read and forward the modified packets.

Rewriting packet headers is an essential component of Layer 3 forwarding; so this feature prepares the packet for Layer 3 forwarding. Another application is IP multicasting where creating multiple versions of the first cell (containing the packet header) in a stream enables transmission to several multicast member ports with minimal processing overhead.

8.6.5 Switched Port Analyzer

There are times when a network manager would need to gather data passing through a switch port to and from a specific network segment or end station. The Switched Port Analyzer (SPAN) feature support in the Catalyst 3550 allows a network manager to designate a particular port (destination port) on the switch to "mirror" activity through specific ports of interest in the system. External sniffers or probes (such as a Cisco SwitchProbe) can be attached to the destination port to gather data passing through the other (source) ports of interest. Remote SPAN (RSPAN) extends SPAN to allow remote monitoring of multiple switches across a network.

9

ARCHITECTURES WITH BUS-BASED SWITCH FABRICS: CASE STUDY— CISCO CATALYST 6500 SERIES SWITCHES WITH SUPERVISOR ENGINE 32

9.1 INTRODUCTION

The Cisco Catalyst 6500 is a family of switch/routers supporting a range of Supervisor Engine and line card options. The older generation of the Cisco Catalyst 6500 supports Supervisor Engines 1A or 2 and two backplanes. One backplane is a 32 Gb/s shared switching bus for interconnecting line cards within the switch/router, and the another backplane allows line cards to interconnect over a high-speed crossbar switch fabric.

The crossbar switch fabric provides each connecting module with a set of distinct high-speed switching paths for data transmission to and data reception from the crossbar switch fabric. This first generation switch fabric (implemented in a stand-alone switch fabric module) provides a total switching capacity of 256 Gb/s.

The newer generation Catalyst 6500 was introduced with the newer high-performing Supervisor Engines 32 and 720 that have advanced features beyond the Supervisor Engines 1A and 2 [CISC6500DS04, CISCCAT6500]. The discussion in this chapter focuses on the architectures of the Catalyst 6500 Series with Supervisor Engine 32. The Supervisor Engine 32 provides connectivity only to the 32 Gb/s shared switching bus and, as a result, supports only line cards that connect to this switching bus. These line card types are the "classic" (also called nonfabric-enabled) and CEF256-based line cards.

Switch/Router Architectures: Shared-Bus and Shared-Memory Based Systems, First Edition. James Aweya.
© 2018 The Institute of Electrical and Electronics Engineers, Inc. Published 2018 by John Wiley & Sons, Inc.

By supporting these line card types, the higher performing Supervisor Engine 32 provides investment protection to users who have already deployed Cisco Catalyst 6500 modules that connect to the 32 Gb/s backplane. The Supervisor Engine 32 protects current 32 Gb/s-based switch investments by supporting all existing "classic" and CEF256-based line cards.

As will be described below, the Supervisor Engine 32 has two uplink options: eight-port Gigabit Ethernet (GbE) Small Form Pluggable (SFP)-based uplinks and two-port 10 GbE XENPAK-based uplinks. The Catalyst 6500 with Supervisor Engine 32 is designed primarily for deployment at the network access layer.

Adopting the architecture categories broadly used to classify the various designs in Chapter 3, the following architectures are covered in this chapter:

- Architectures with bus-based switch fabrics and centralized forwarding engines (see Figure 9.1):
  - Architectures with "Classic" line cards.
  - Architectures with CEF256 line cards (optional Distributed Forwarding Card (DFC) not installed).

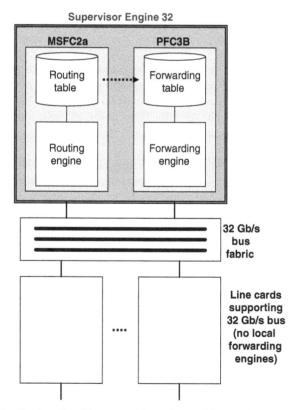

FIGURE 9.1 Bus-based architecture with routing and forwarding engines in separate centralized processors.

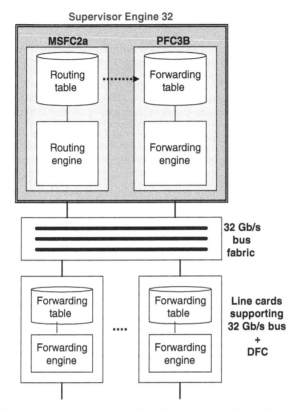

FIGURE 9.2 Bus-based architecture with fully distributed forwarding engines in line cards.

- Architectures with bus-based switch fabrics and distributed forwarding engines (see Figure 9.2):
  - Architectures with CEF256 line cards with optional DFC installed.

9.2 CISCO CATALYST 6500 32 Gb/s SHARED SWITCHING BUS

The 32 Gb/s switching bus allows all the modules connected to it to share the common bandwidth available for both data transmission and data reception. As described in Chapter 7, the shared bus consists of three (sub-)buses, each playing a specific role in the data forwarding operation in the system. These buses are the Data Bus (DBus), Results Bus (RBus), and the Control Bus (CBus).

The DBus is the main system bus that carries all end-user data transmitted and received between modules. It has a bandwidth of 32 Gb/s (i.e., 2×256 bits wide $\times 62.5$ MHz clock speed). The "32 Gb/s" in the name of the switching bus comes from this data transfer rate. The DBus is a shared bus, so to transmit a packet, a line card must arbitrate for access to the DBus by submitting a transmit request to a master arbitration mechanism that is located on the Supervisor Engine (or primary Supervisor Engine if a redundant Supervisor Engine is installed).

If the DBus is not busy, the master arbitration mechanism grants access permitting the line card to transmit the packet on the DBus. With this bus access permission, the line card transmits the packet over the DBus to the Supervisor Engine. During the packet transfer over the DBus, all connected line cards will sense the packet being transmitted and capture a copy of the packet into their local buffers.

The Supervisor Engine uses the RBus to forward the forwarding instructions (obtained after forwarding table lookup) to each of the attached line cards. The forwarding instruction sent by the Supervisor Engine to each line card is either a drop or forward action. A drop action means a line card should flush the packet from its buffers, and a forward action means the packet should be sent out a port to its destination. The CBus (or Ethernet out-of-band channel (EOBC)) is the bus that carries control information between the line cards and the control and management entity on the Supervisor Engine.

9.2.1 Main Features of the Catalyst 6500 Shared Bus

The Cisco Catalyst 6500 shared switching bus employs two methods to achieve improved performance over the traditional shared bus: pipelining and burst mode.

9.2.1.1 Pipelining Mode The traditional implementation of the shared bus allows a single frame transmission over the shared bus at any given point in time. Let us consider the situation where the system employs a traditional shared bus. The Supervisor Engine receives a packet from a line card and performs a lookup into its local forwarding table to determine which line card port the packet should be forwarded to. It sends the result of the forwarding table lookup to all ports connected to the shared bus over the RBus.

In the traditional implementation, while the table lookup is being performed, no subsequent packets are sent over the bus. This means there are some idle times in data transfers over the bus resulting in suboptimal use of the bus – Bus utilization is not maximized.

The Catalyst 6500 employs pipelining to allow ports to transmit up to 31 frames across the shared bus (to be pipelined at the Supervisor Engine for lookup operation) before a lookup result is transmitted via the RBus. If it happens that there is a 32nd packet to be transmitted, it will be held locally at the transmitting port until the port receives a result over the RBus. Pipelining allows the system to reduce the idle times that would have been experienced in the traditional bus implementation and also provides improvements in the overall utilization of the shared bus architecture.

9.2.1.2 Burst Mode Another concern in the use of traditional shared bus is that the bus usage could unfairly favor ports transmitting larger frames. Let us consider, for example, two ports that are requesting access to the shared bus. Let us assume that Port A is transmitting 512 byte frames and Port B is transmitting 1518 byte frames. Port B would gain an unfair bus usage advantage over Port A when it sends a sequence of frames over a period of time because it consumes relatively more bandwidth in the process. The Catalyst 6500 uses the burst mode feature to mitigate this kind of unfairness.

To implement the burst mode feature, the port ASIC (which handles access to the shared bus) maintains a count of the number of bytes it has transmitted and compares this with a locally configured threshold. If the byte count is below the threshold, then a packet waiting to be transmitted can be forwarded. If the byte count exceeds the threshold, then the port ASIC stops transmitting frames and bus access is removed for this port (done by the master arbitration mechanism in the Supervisor Engine). The threshold is computed by the port using a number of local variables extracted from the system (see related discussion in Chapter 7) in order to ensure fair distribution of bus bandwidth.

9.3 SUPERVISOR ENGINE 32

The Supervisor Engine is the main module in the Catalyst 6500 responsible for all centralized control plane and data plane operations. The control plane is responsible for running the routing protocols and generating the routing table that contains the network topology information (location of each destination IP address in the network). Each destination IP address in the routing table is associated with a next hop IP address, which represents the next closest Layer 3 device or router to the final destination. The contents of the routing table are distilled into a much compact and simple table called the forwarding table.

The data plane is responsible for the operations that are actually performed on a packet in order to forward it to the next hop. These operations involve performing a forwarding table lookup to determine the next hop address and egress interface, decrementing the IP TTL, recomputing the IP checksum, rewriting the appropriate source and destination Ethernet MAC addresses in the frame, recomputing the Ethernet checksum, and then forwarding the packet out the appropriate egress interface to the next hop. Control plane functions are typically handled in software, whereas data plane functions are simple enough to be implemented in hardware, if required.

9.3.1 Supervisor Engine 32 Architecture

The Supervisor Engine 32 has connectivity only to the 32 Gb/s shared bus and supports packet forwarding of up to 15 Mpps [CISCSUPENG32]. Unlike the Supervisor Engines 2 and 720, Supervisor Engine 32 does not provide connectivity to a crossbar switch fabric, as illustrated in Figures 9.3 and 9.4. As shown in these figures, Supervisor Engine 32 supports the PFC3B and MSFC2a as a default configuration.

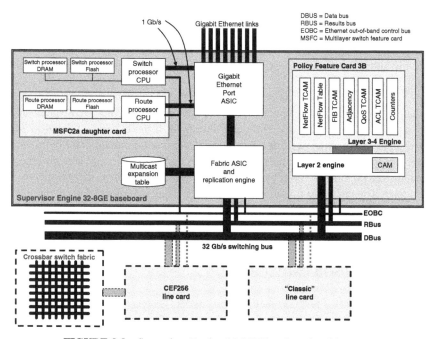

FIGURE 9.3   Supervisor Engine 32-8GE baseboard architecture.

The Supervisor Engine 32 comes in two versions:

- **Supervisor Engine 32-8GE:** This Supervisor Engine option supports 8 GbE Small Form-Factor Pluggable uplink ports (see Figure 9.3).
- **Supervisor Engine 32-10GE:** This Supervisor Engine option supports 2×10 GbE uplink ports (see Figure 9.4).

These Supervisor Engines also support an additional 10/100/1000TX front port and two USB ports on the front panel (type A and type B USB port). The type A USB port, which is designated for host use, can be used to plug in devices such as a laptop or PC. The type B is designated as a device port and can be used for attaching devices such as a Flash memory key.

A chassis that supports redundancy with two Supervisor Engine 32-8GE modules provides a total of 18 active Gigabit Ethernet ports (i.e., $2 \times (8 + 1)$ ports) to the user, where all ports on both the primary and secondary Supervisor Engines are active.

9.3.2   Multilayer Switch Feature Card 2a (MSFC2a)

The MSFC2a provides Layer 3 control plane functionality that enables the Supervisor Engine 32 to function as a full-fledged Layer 3 device. Without the

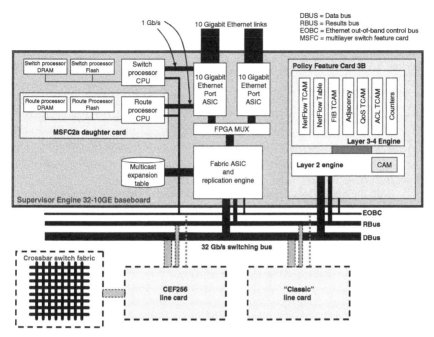

FIGURE 9.4   Supervisor Engine 32-10GE baseboard architecture.

MSFC2a, the Supervisor Engine 32 will function purely as a Layer 2 device. Forwarding using network topology-based forwarding tables and optimized lookup algorithms (called CEF (Cisco Express Forwarding)) is the forwarding architecture implemented in the Supervisor Engine 32. The MSFC2a and the MSFC2 used in the Supervisor Engine 2 (see Chapter 7) are functionality equivalent, except the MSFC2a uses a bigger DRAM.

In the Supervisor Engine 32, a Switch Processor CPU, as shown in Figures 9.3 and 9.4, is responsible for running all the Layer 2 control plane protocols, such as Spanning Tree Protocol (STP), IEEE 802.1AB Link Layer Discovery Protocol (LLDP), and VLAN Trunking Protocol (VTP). The Switch Processor is allocated its own (upgradeable) DRAM and nonvolatile RAM (NVRAM).

A route processor CPU on the MSFC2a (Figures 9.3 and 9.4) is responsible for running the Layer 3 routing protocols and ICMP, carrying out address resolution functions to map IP addresses to Layer 2 addresses, initializing and managing switched virtual interfaces (SVIs), and running and configuration of the Cisco IOS Software. An Ethernet out-of-band control bus (a full duplex, 1 Gb/s in-band connection) shown in Figures 9.3 and 9.4) enables the MSFC2a to communicate and exchange information with other entities on the Supervisor Engine 32 baseboard.

The MSFC2a communicates with its Layer 3 peers in a network (via the configured routing protocols (Open Shortest Path First (OSPF), Enhanced

Interior Gateway Routing Protocol (EIGRP), border Gateway Protocol (BGP), etc.) and generates routing information about the network topology (which is maintained in the routing table). The MSFC2a then distills this routing information to generate a more compact forwarding table or FIB, which is then forwarded to the PFC3B.

The PFC3B stores the forwarding table in a FIB ternary content addressable memory (TCAM). As shown in Figures 9.3 and 9.4, the FIB TCAM is implemented on the PFC3B daughter card and is a very high-speed memory that allows the PFC3B to perform fast forwarding table lookups during packet forwarding. The Layer 3 features that are supported on a Supervisor Engine 32 are described in more detail in [CISCSUPENG32].

As with the Supervisor Engine 720, the Supervisor Engine 32 also implements hardware counters, registers, and control plane policing (CoPP) (of control plane traffic) to limit the effect of denial of service (DoS) attacks on the control plane. The hardware-based control plane policing allows a control plane quality of service (QoS) policy to be applied to the Supervisor Engine 32 in order to limit the total amount of traffic that is sent to its control plane.

9.3.3   Policy Feature Card 3B (PFC3B)

The PFC3B (Figures 9.3 and 9.4) supports hardware-based features that allow the Supervisor Engine 32 to perform more enhanced QoS and security operations. For example, to secure and prioritize data, the PFC3B provides hardware support for security and QoS-based ACLs using Layer 2, 3, and 4 classification criteria. The PFC3B also allows the Supervisor Engine 32 to support new hardware accelerated features such as ACL hit counters, port access control lists (PACLs), Enhanced Remote Switched Port Analyzer (ERSPAN), CPU rate limiters, IP Source Guard, and NetFlow capacities:

- **ACL Hit Counters:** ACL hit counters allow a network administrator to monitor how many times (i.e., hits) a specific access control entry (ACE) within an ACL has been applied as traffic pass through the device port or interface. The ACE hit patterns (number of hits) provide the user with additional information that can be used to fine-tune the ACLs to be more effective for the traffic the ACLs are applied to.
- **Port ACLs (PACLs):** A PACL is an ACL that can be applied at a single port on a Layer 2 switch within a VLAN (i.e., physical switch port or trunk port that belongs to a specific VLAN). It provides a functionality similar to a VLAN ACL (VACL), but unlike a VACL (which when applied, extends and covers an entire VLAN), the PACL is applied only to a single port within a VLAN. The PACL can be applied at the ingress of a switch port to screen ingress traffic and is processed before any VACLs that may be applied to the switch port.

- **Enhanced Remote SPAN (ERSPAN):** The switched port analyzer (SPAN) feature (also referred to as port mirroring or port monitoring) is a feature that can be applied to a switch port to copy or sample traffic to be examined and analyzed by a network traffic analyzer such as a Remote Monitoring (RMON) probe. ERSPAN allows a switch to forward a copy of traffic passing through a switch port to a destination SPAN port that may be located in another network reachable over multiple Layer 3 hops. For example, ERSPAN can be applied on a switch port located in one IP subnet while the destination SPAN port is located in another subnet. ERSPAN uses the tunneling protocol, generic routing encapsulation (GRE), to carry the traffic over the Layer 3 network to the destination SPAN port.

- **CPU Rate Limiting:** The control plane provides a critical set of functions to a switch, switch/router, or router. CPU rate limiters are rate-limiting mechanisms that can be applied to different traffic types sent to the control plane of the switch, switch/router, or router. The CPU rate limiters in Supervisor Engine 32 are designed to protect the performance of the Layer 3 control plane (or route processor) from being overwhelmed or overloaded with unnecessary or DoS traffic. From a performance perspective, implementing CPU rate limiters in hardware further strengthens the Supervisor Engine 32 from high-speed attacks that can compromise the performance of the system. The Layer 2, 3, and application rate limiters as well as unicast and multicast rate limiters are listed in [CISCSUPENG32].

- **Bidirectional Protocol-Independent Multicast (BIDIR-PIM):** BIDIR-PIM allows the network nodes involved in multicast traffic distribution to construct bidirectional multicast distribution trees, which support bidirectional traffic flow. BIDIR-PIM provides an alternative multicast distribution model to the other PIM distribution models (PIM Dense Mode (PIM-DM), PIM Sparse Mode (PIM-SM), PIM Source-Specific Multicast (PIM-SSM)) and is designed to support many-to-many multicast communications within a single PIM domain. BIDIR-PIM also reduces or lessens the amount of PIM forwarding state information that a multicast router must maintain (in its multicast forwarding table), a feature that is particularly important during many-to-many multicast transfers with many and widely spread out data senders and receivers. The BIDIR-PIM shared trees provide a valuable feature in that many multicast sources can transmit on the same shared tree without the multicast routers having to explicitly maintain state for each source. This feature also provides the added benefit of reduced processing load on the Supervisor Engine's CPU and memory.

- **IP Source Guard:** Spoofing attacks occur when a client that is unknown or untrusted floods a network with malicious packets that are intended to harm the operations of the network. Often the clients in these attacks utilize source IP address spoofing to hide the true source of the attack. Hackers often use spoofed packets to gain access into a network by changing their true source IP address to one that is recognized by the network as a genuine internal or secure address. IP Source Guard is a method that can be used to provide protection against spoofed packets. Using Dynamic Host Configuration Protocol (DHCP) snooping, IP Source Guard snoops on DHCP requests and constructs a dynamic PACL (Port ACL) that is applied to incoming traffic to deny all packets that do not match the assigned DHCP address. The PACL is applied at an interface of the device running the IP Source Guard feature.

Other QoS services supported on the PFC3B include ingress traffic policing and classification of incoming data. This allows the rewrite of IEEE 802.1p class of service (CoS) bits in the Ethernet header and IP Precedence/DSCP priority bits in the IPv4 header.

9.3.4 Supervisor Engine 32 as a "Classic" Module

The Supervisor Engine 32 is also referred to as a "Classic" module (see Chapter 7) because it supports connectivity only to the "Classic" 32 Gb/s shared switching bus that allows it to communicate with other line cards connected to the bus. The Supervisor Engine 32 also has no built-in crossbar switch fabric (as in other Supervisor Engines like Supervisor Engine 720), nor does it support connectivity to a separate crossbar switch fabric module (like in Supervisor Engine 2).

The support of only the Classic (32 Gb/s) shared bus thus dictate the type of line cards that can operate with the Supervisor Engine 32. Line cards that do not support connectivity over the Classic 32 Gb/s shared bus cannot be used with the Supervisor Engine 32.

A full list of the line card architectures supported with the Supervisor Engine 32 is given in [CISCSUPENG32]. As explained below, the Supervisor Engine 32 supports both the CEF256 and Classic line card architectures. Both of these line cards have a connector on the line card that provides connectivity to the Classic 32 Gb/s bus.

The 32 Gb/s shared bus allows all ports connected (on both the Supervisor Engine 32 and line cards) to exchange data. The DBus is 256 bits wide and is clocked at 62.5 MHz, which yields bandwidth of 16 Gb/s. The RBus also operates at 62.5 MHz and is 64 bits wide.

9.3.5 Fabric ASIC and Replication Engine

The Supervisor Engine 32 baseboard (Figures 9.3 and 9.4) has a number of onboard application-specific integrated circuits (ASICs) that enable the support of Layer 2, 3, and 4 services and also serves as an interface to the 32 Gb/s shared switching bus. One ASIC is used to connect the Supervisor Engine to the 32 Gb/s shared switching bus. This specialized ASIC (referred to as the Fabric ASIC and Replication Engine) is also used for multicast packet replication in the Supervisor Engine 32 and supports the SPAN functionality used for port mirroring.

The Supervisor Engine 32 via the Fabric ASIC and Replication Engine supports only ingress replication mode of multicast packets. As shown in Figures 9.3 and 9.4, the Fabric ASIC and Replication Engine also provides an interface to the multicast expansion table (MET), which supplies the Supervisor Engine 32 with the relevant information regarding the multicast group membership it serves. Another onboard port ASIC holds the port interface logic that provides connectivity to the 9 GbE ports (Figure 9.3) or the two 10 GbE ports (Figure 9.4).

9.4 CATALYST 6500 LINE CARDS SUPPORTED BY SUPERVISOR ENGINE 32

The Cisco Catalyst 6500 supports a number of line card types with different physical media types and speed options. These line cards are designed with a range of features to allow the Catalyst 6500 to meet the needs of deployment in the access, distribution, and core layers of a network. A line card slot may provide a connection to the 32 Gb/s shared bus and in some designs, another connection to a crossbar switch fabric if either a Supervisor Engine 2 or 720 is present.

The Catalyst 6500 supports four general line card types, that is, the Classic, CEF256, CEF720, dCEF256, and dCEF720 line cards [CISCCAT6500]. All of these line cards can interoperate and communicate with each other when installed in the same chassis as long as the relevant fabric connections are present in the chassis. The Catalyst 6500 with Supervisor Engine 32 supports only the line cards that have connectivity to the 32 Gb/s shared bus – the Classic and CEF256 line cards.

9.4.1 Classic Line Card Architecture

The Classic line card (also called the nonfabric-enabled line card (see discussion in Chapter 7)) supports connectivity only to the 32 Gb/s shared switching bus. It has a shared bus connection only and no connection to a stand-alone or Supervisor Engine integrated crossbar switch fabric. Furthermore, it does not support packet forwarding locally in the line card (i.e., distributed forwarding).

All generations and versions of the Supervisor Engines, from the Supervisor Engine 1A through to the newer Supervisor Engine 720-3BXL, support the Classic line cards. A Classic line card when installed in a Cisco Catalyst 6500 chassis does not allow the line cards to operate in compact (switching) mode (see bus switching modes discussion below). Thus, with the presence of this line card, the centralized forwarding rate of the PFC3B reaches only 15 Mpps.

9.4.2 CEF256 Line Card Architecture

Figure 9.5 shows the architecture of the CEF256 line card. The CEF256 line card is a fabric-enabled line card and supports one connection to the 32 Gb/s shared switching bus and another connection to the crossbar switch fabric [CISCCAT6500]. The connection to the crossbar switch fabric is a single 8 Gb/s fabric channel. The line card also supports a single internal 16 Gb/s local shared switching bus over which local packets are forwarded. The 16 Gb/s local shared switching bus has a similar function and operation as the main chassis 32 Gb/s shared bus. The chassis 32 Gb/s shared bus is the main bus that connects all shared switching bus capable line cards (i.e., the nonfabric-enabled and fabric-enabled line cards) in the Cisco Catalyst 6500 chassis.

The 16 Gb/s local switching bus on the CEF256 line card is utilized for forwarding packets that have port destinations local within the line card. Using this bus, a packet that is to be forwarded locally (utilizing an optional DFC or DFC3a to determine the forwarding destination) avoids being sent over the 32 Gb/s shared bus or the crossbar switch fabric. This local forwarding capability reduces the overall latency of forwarding packets and frees up the chassis 32 Gb/s shared bus or crossbar switch fabric capacity for those line cards that cannot forward packets locally.

As shown in Figure 9.5, the CEF256 line card supports internally a fabric Interface ASIC, which serves as the interface between the local ports on the line card and other modules connected to the crossbar switch fabric. The fabric Interface ASIC also allows line card to connect to the 32 Gb/s shared switching bus.

The CEF256 line cards will use the crossbar switch fabric for forwarding packets to other modules when it is installed in a chassis with a Supervisor Engine 720.

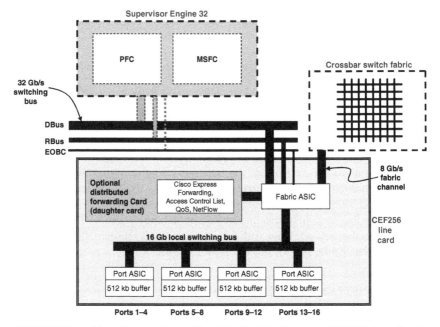

FIGURE 9.5   Cisco Express Forwarding Line Card Architecture (CEF256 Line Card).

However, if a Supervisor Engine 32 is installed, the system will fall back to using the 32 Gb/s shared switching bus since the Supervisor Engine 32 supports connectivity only to the 32 Gb/s shared switching bus.

9.5 CISCO CATALYST 6500 32 Gb/s SHARED SWITCHING BUS MODES

We describe here the three switching modes used by the 32 Gb/s shared switching bus and fabric interface ASICs in the CEF256 and CEF720 line cards [CISCCAT6500]. The switching modes define the format of the internal packet header or forwarding tag used to transfer data across the DBus (of the 32 Gb/s shared switching bus) and also communicate with other CEF256 and CEF720 modules in the chassis. These switching modes do not apply to line cards that support Distributed Forwarding Card (DFC) feature.

9.5.1 Flow-Through Mode

The Flow-Through mode of operation is used by the CEF256 (fabric-enabled) line cards when a crossbar switch fabric is not present in the chassis. This mode enables CEF256 line cards to operate in the system and over only the 32 Gb/s shared bus as if they were Classic (nonfabric-enabled) line cards. This mode does not apply to the dCEF256, CEF720, and dCEF720 line cards because they do not support connectivity to the 32 Gb/s shared bus.

In this mode, the whole (original) packet (i.e., the original packet header and data) is forwarded by the CEF256 line card over the 32 Gb/s shared bus to the Supervisor Engine for forwarding table lookup and forwarding to the destination port. In the flow-through mode, the Catalyst 6500 achieves a centralized forwarding rate of up to 15 Mpps.

9.5.2 Compact Mode

For a system to operate in the compact mode, the system must support a crossbar switch fabric in addition to the 32 Gb/s shared switching bus. The crossbar switch fabric can be realized in the form of a stand-alone switch fabric module (installed in a chassis slot) or a Supervisor Engine 720 (which has an integrated crossbar switch fabric). The line cards in the chassis must all be fabric enabled (i.e., CEF256) for the system to run in compact mode.

Classic line cards (which do not have a crossbar switch fabric connection), when installed in the chassis, will not allow the system to operate in compact mode. Note that the dCEF256, CEF720, and dCEF720 line cards do not have connections to the 32 Gb/s shared bus.

In compact mode, a line card will send only the (original) packet header over the DBus of the 32 Gb/s shared bus to the Supervisor Engine for processing. To conserve DBus bandwidth and to allow for faster header transmission, the original packet header is compressed before to being transmitted on the DBus. The line card transmits the data portion of the packet over the crossbar switch fabric channels to the destination port. In this mode, the system achieves (independent of packet size) a centralized forwarding rate of up to 30 Mpps.

9.5.3   Truncated Mode

The truncated mode is used when the chassis has the following three module types present: a crossbar switch fabric, CEF256 and/or CEF720 line cards, and Classic line cards. When operating in this mode, the Classic line cards will forward over the DBus of the 32 Gb/s shared bus to the Supervisor Engine, the header, plus the data portion of the (original) packet. The CEF256 and CEF720 line cards, on the other hand, will forward the packet header over the DBus and the data portion over the crossbar switch fabric.

In the truncated mode, the system achieves a centralized forwarding rate up to 15 Mpps. Furthermore, in this mode, because the CEF256 and CEF720 line cards transmit the data portion of the packet over the crossbar switch fabric, the overall aggregate bandwidth achieved can be higher than the 32 Gb/s shared switching bus capacity. However, the performance of line cards that have the DFC feature is not affected by the truncated mode – the forwarding performance stays the same and does not change regardless of the line card mix in the system.

9.6   SUPERVISOR ENGINE 32 QoS FEATURES

We review in this section the queue structures and QoS features on the uplink ports (Figures 9.3 and 9.4) of the Supervisor Engine 32 [CISCSUPENG32].

9.6.1   Uplink Port Queues and Buffering

The transmit side of each Gigabit Ethernet uplink port (see Figure 9.3) is assigned a single strict priority queue and three normal (lower priority) queues. Each of these normal transmit queues supports eight queue fill thresholds, which can be used with a port congestion management algorithm for congestion control. The receive side is assigned two normal queues, each with eight queue fill thresholds for congestion management. There receive side has no strict priority queue.

Each of the Ethernet uplink ports on the Supervisor Engine 32 Gigabit (Figure 9.3) is allocated 9.5 MB of per port buffering. The 10 GbE ports (Figure 9.4), on the other hand, are assigned 100 MB of per port buffering. The provision of large per port buffering is of particular importance when the switch is operating in networks that carry bursty applications or high data volume applications (e.g., long flow TCP sessions, network video, etc.). With large buffering per port, these applications can use the extra buffering should the data transfers become very bursty.

9.6.2 DSCP Transparency

Both the Supervisor Engine 32 and Supervisor Engine 720 support a feature called differentiated services code point (DSCP) transparency. DSCP transparency is a feature that allows the switch to maintain the integrity of the DSCP bits carried in a packet as it transits the switch. Let us consider the situation where a packet arrives on a switch port carrying traffic that is not trusted (an untrusted port) and the switch assigns a lower class-of-service (CoS) value to the packet.

From this incoming CoS value, the switch derives an internal priority value that is used to write the DSCP bits on egress. DSCP transparency prevents this situation, and similar ones, by not allowing the switch to use the internal priority to derive the egress DSCP value. Instead, the switch will simply write the ingress DSCP value on egress.

When DSCP transparency is not used, the DSCP field in an incoming packet will be modified by the switch, and the DSCP field in the outgoing packet will be modified based on the port QoS settings that may include the policing and marking policies configured, port trust level setting, and the DSCP-to-DSCP mutation map configured at the port.

On the other hand, if DSCP transparency is used, the DSCP field in the incoming packet will not be modified by the switch, and the DSCP field in the outgoing packet will not be modified and stays unchanged – The value is the same as that in the incoming packet. It is worth noting that regardless of whether DSCP transparency is used or not, the switch will still use an internal DSCP value for the packet, which it uses for internal traffic processing to generate a CoS value that reflects the priority of the traffic. The internal DSCP value is also used by the switch to select an egress queue and queue fill threshold for the outgoing packet.

9.6.3 Traffic Scheduling Mechanisms

Two important scheduling mechanisms that can be used on the Supervisor Engine 32 GbE uplink ports (Figure 9.3) are the shaped round-robin (SRR) and deficit weighted round-robin (DWRR) algorithms. SRR allows the maximum amount of bandwidth that each queue is allowed to use to be defined. SRR like DWRR requires a scheduling weight to be configured for each of the queues, but the weight values are used differently in SRR.

After the scheduling weights are assigned to all queues, the SRR algorithm normalizes the total of the weights to 1 (or equivalently, 100%). Then a maximum bandwidth value is derived from the normalized values and assigned to each queue. The flow of data out of the queue will then be shaped to not exceed this (maximum) bandwidth value. But unlike DWRR, each queue that is shaped will not be allowed to exceed the maximum bandwidth value computed from the normalized weights. With SRR, traffic in excess of the maximum bandwidth value will be buffered and scheduled resulting in the traffic appearing to have a smoothing output over a given period of time.

The DWRR algorithm, on the other hand, aims to provide a fairer allocation of bandwidth between the queues than when the ordinary weighted round-robin (WRR) is used. The weights in DWRR determine how much bandwidth each queue is allowed to use, but, in addition, the algorithm maintains a measure or count of excess bandwidth each queue has used.

To understand the DWRR algorithm, let us consider, for example, a queue that has used up all but 500 bytes of its allocation, but has another packet in the queue that is 1500 bytes in size. When this 1500 packet is scheduled, the queue has consumed 1000 bytes of bandwidth in excess of its allocation on that scheduling round. The DWRR algorithm works by recognizing that an extra 1000 bytes has been used and deducts this (excess bytes) from the queue's bandwidth allocation in the next scheduling round. When the operation of the DWRR algorithm is viewed over a period of time, all the queues will on average be served closer to their allocated portion of the overall bandwidth.

9.7 PACKET FLOW THROUGH SUPERVISOR ENGINE 32

This section describes how a packet is forwarded through the Supervisor Engine 32. The steps involved in forwarding packets over the shared bus are described below and are also marked in Figures 9.6 and 9.7.

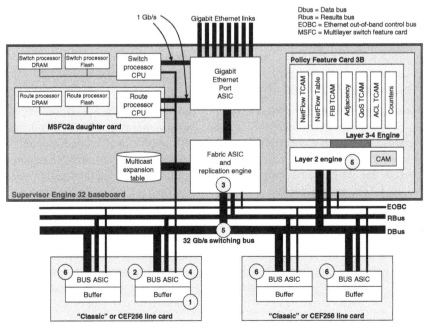

FIGURE 9.6   Packet Flow through the Supervisor Engine 32 – Steps 1–6 of the packet flow.

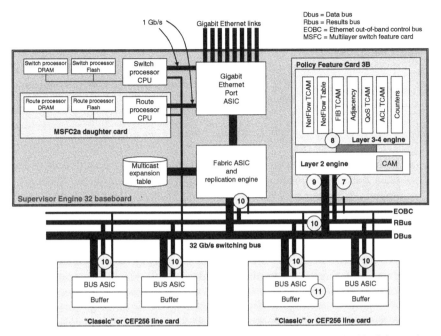

FIGURE 9.7　Packet flow through the Supervisor Engine 32 – Steps 7–11 of the packet flow.

Step 1: Packet Enters an Input Port on the Switch from the Network

- A packet from the network enters a port on a Classic or CEF256 line card and is temporarily stored in an input buffer.
- Based on the arriving packet header information, the port BUS ASIC constructs an internal tag or header carrying information that the centralized Supervisor Engine 32 forwarding engine will use to forward the packet.
- This information will be used by the Supervisor Engine 32 to perform lookup in its forwarding table and also apply any necessary QoS and security policies. All the necessary ingress QoS policies can also be applied here if configured.

Step 2:– Ingress Port Arbitrates for Access to 32 Gb/s Shared Switching Bus

- The port BUS ASIC on the ingress line card arbitrates for access to the 32 Gb/s shared bus to enable it transmit the packet to the Supervisor Engine 32.
- Each line card has a local arbitration mechanism that allows the line card to communicate and send bus access requests to the central arbitration mechanism located on the Supervisor Engine 32.

Step 3:– Access Granted to 32 Gb/s Shared Bus

- If the 32 Gb/s shared bus is idle, the central arbitration mechanism on the Supervisor Engine 32 will grant access to the bus (by forwarding a grant or permit message to the local arbitration mechanism on the line card indicating that it is allowed to transmit).

Step 4: Ingress Port Forwards Packet over the 32 Gb/s Shared Bus

- When the local arbitration mechanism on the ingress line card receives the grant message from the central arbitration mechanism on the Supervisor Engine 32, the port BUS ASIC transmits the packet on the 32 Gb/s shared bus.

Step 5: Packet Received by the Supervisor Engine 32

- The port BUS ASIC forwards the packet over the 32 Gb/s shared bus to the Supervisor Engine 32 and is received by the Layer 2 forwarding engine located on the PFC3B.

Step 6: All Line Cards Connected to the 32 Gb/s Shared See the Transmitted Packet

- Given that the 32 Gb/s shared bus is a shared medium, all other line cards connected to this shared bus will sense the transmitted packet and copy the packet temporarily in their transmit buffers.
- This packet will be stored in the transmit buffers of the line cards until the Supervisor Engine 32 informs the line cards to either drop or forward the packet out to the network.

Step 7: Layer 2 Forwarding Engine in the PFC3B Processes the Packet

- The Layer 2 forwarding engine on the PFC3B receives the packet and performs a lookup in its Layer 2 forwarding table using the destination MAC address in the packet.
- After this Layer 2 operation, the packet is passed on to the Layer 3 forwarding engine (on the PFC3B) for further processing.

Step 8: Layer 3 Forwarding Engine in the PFC3B Processes the Packet

- The Layer 3 forwarding engine on the PFC3B then executes a number of tasks in parallel in order to forward the packet.
- Upon system start-up and initialization, the MSFC2a continues to run the routing protocols and populate the FIB TCAM located on the PFC3B that

maintains a view of the network topology. The FIB is generated from the master routing tables created by the routing protocols running on the MSFC2a.

- The PFC3B Layer 3 forwarding engine would perform a FIB lookup if the packet forwarding operation demands a Layer 3 forwarding operation.

- In parallel to the Layer 3 lookup process, lookups are also performed in the QoS TCAM and security ACL TCAM to determine if any of the QoS and ACLs configured need to be applied to this packet.

- If necessary, the PFC3B will also update NetFlow statistics for the flow that this packet belongs to.

Step 9: PFC3B Generates Lookup Results

- The PFC3B assembles the results of all lookup operations that contain the following key information:
  - Instructions to the egress line cards to either drop or forward the packet.
  - MAC rewrite information to be used by the egress line cards to modify the Layer 2 MAC destination address (corresponding to the next hop IP address) so that the packet can be sent to its correct next hop node
  - QoS information instructing the egress line cards to store the packet into the correct output port queue and any rewrite information needed for modifying outgoing DSCP values

Step 10: PFC3B Forwards Lookup Results over the RBus to Egress Line Cards

- The PFC3B forwards the result of the lookup operation over the RBus to all destination line cards and ports.

Step 11: Destination Ports Receive Lookup Results from the PFC3B

- The line cards and destination ports receive (over the RBus) the results information from the PFC3B and use this information to construct the Ethernet frame header for the outgoing packet.

- The reconstructed packet is extracted from the destination port's local transmit buffer and transmitted out the egress interface to the network.

10

ARCHITECTURES WITH SHARED-MEMORY-BASED SWITCH FABRICS: CASE STUDY—CISCO CATALYST 8500 CSR SERIES

10.1 INTRODUCTION

The Catalyst 8500 campus switch/routers (CSRs) are modular networking devices that provide wire speed Layer 2 and Layer 3 packet forwarding and services [CISCCAT8500]. The Catalyst 8500 family of devices comprises the Catalyst 8510 and 8540 switch/routers. This family of high-speed switch/routers is targeted for campus or enterprise backbones.

The Catalyst 8500 family of devices support both IP and IPX routing standards but the focus of this chapter will be on the IP routing features. The IPX protocol has been deprecated and is no more in use. Based on architecture categories described in Chapter 3, the architectures discussed here fall under "architectures with shared-memory-based switch fabrics and distributed forwarding engines" (see Figure 10.1).

10.2 MAIN ARCHITECTURAL FEATURES OF THE CATALYST 8500 SERIES

The key features of the Catalyst 8500 CSR include wire-speed Layer 3 IP unicast and multicast forwarding over 10/100 Mb/s Ethernet and Gigabit Ethernet (GbE) interfaces. The switch/router supports the configuration of virtual LANs (VLANs) between switches via the IEEE 802.1Q standard and Cisco Inter-Switch Link (ISL)

Switch/Router Architectures: Shared-Bus and Shared-Memory Based Systems, First Edition. James Aweya.
© 2018 The Institute of Electrical and Electronics Engineers, Inc. Published 2018 by John Wiley & Sons, Inc.

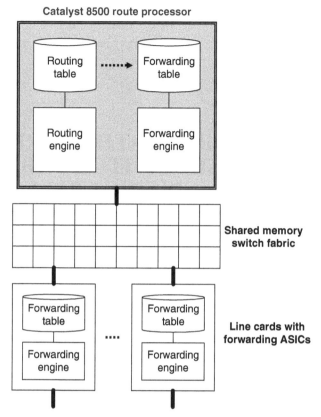

FIGURE 10.1   Architectures with shared-memory-based switch fabrics and distributed forwarding engines.

trunking protocol (a Cisco proprietary protocol for encapsulating Ethernet frames with VLAN information).

The switch/router also supports a number of quality of service (QoS) features, including four priority queues per port, and flow classification and priority queuing based on IP Precedence bit settings. The main architectural features of the Catalyst 8510 and 8540 are discussed in this chapter.

10.2.1   Catalyst 8510

The Catalyst 8510 family supports wire-speed IP packet nonblocking forwarding on all ports and forwarding rates of up to 6 million packets per second (Mpps). The Catalyst 8510 employs a five-slot modular chassis that can carry up to 32 10/100 Mb/s Ethernet ports or 4 GbE ports (that can be used for uplink connectivity to a network or to servers). The modular chassis also supports two fault-tolerant, load-sharing power supplies (primary and optional secondary).

The Catalyst 8510 switch/router supports a 10 Gb/s full nonblocking shared memory switch fabric for both Layer 2 and Layer 3 forwarding. This shared memory switch fabric allows for an aggregate packet throughput of 6 Mpps. The central slot in the five-slot chassis is dedicated to the shared memory switch fabric and a high-performance Switch Route Processor (SRP) (a RISC processor).

Furthermore, the Catalyst 8510 supports two line card types. One has eight 10/ 100 Mb/s Ethernet ports with Category 5 copper cable and 8P8C (8 position 8 contact), also called T568A/T568B) connectors (but commonly referred to as RJ-45). The second has eight 100BASEFX Ethernet ports with fiber-optic cable and SC fiber connectors.

The SRP module runs the Layer 2 and Layer 3 protocols that provide required intelligence in the Catalyst 8510 for switching and routing packets. The SRP interfaces to each port in the switch/router via the shared memory switch fabric. The switching and routing protocols SRP are implemented as part of the Cisco Internetwork Operating System (IOS) software commonly used in Cisco switching and routing devices. The SRP is responsible for running the routing protocols including the multicast protocols, and the generation and maintenance of the distributed forwarding tables that reside in the line cards.

The SRP also supports SNMP agents and MIBs used for the management of the switch/router. Other features implemented in the SRP are advanced packet classification and management applications used for traffic management. The SRP is carried in the middle slot in the five-slot chassis, while the remaining four slots are used for line card modules. In addition to supporting redundant power supply modules, the Catalyst 8510 supports field replaceable (i.e., hot-swappable) fan trays while the switch/router is operational, thus reducing the mean time to repair.

The SRP module supports two PCMCIA Type II slots into which a variety of Flash EPROM modules can be fitted. These EPROMs introduce 8–20 MB of additional memory to the SRP. The EPROM modules allow the SRP to support larger Cisco IOS software code images as the IOS software is updated and grows. The Flash EPROMs can also be used to program standard configuration parameters for the Catalyst 8510. The EPROMs, however, are not required for the running mode operation of the switch/router, but mostly as a boot EPROM.

10.2.2   Catalyst 8540

The Catalyst 8540 switch/router has a 13-slot chassis and a 40 Gb/s nonblocking shared memory switch fabric. This switch/router can forward packets at rates up to 24 Mpps. Two switch fabric modules are required to hold the shared memory and allow for the transporting of packets from one switch/router interface to another. Similar to the Catalyst 8510, the Catalyst 8540 also supports Layer 2 and Layer 3 forwarding of packets and other IP services.

The Catalyst 8540 also supports two line card types. One has 16 10/100 Mb/s fast Ethernet ports over copper and the other 16 100Base-FX ports over fiber. Each line card can have either 16,000 or 64,000 forwarding table entries. This translates into

16,000 or 64,000 Layer 2 MAC addresses or IP addresses, or a combination of both address types. These addresses can be stored locally in a forwarding table within a line card to be used for Layer 2 and 3 packet forwarding. The Catalyst 8540 also supports a number of system redundancy features, which include redundant switch fabrics, system processors, and power supply modules.

The Layer 2 and 3 routing information required for forwarding packets in the Catalyst 8540 is provided by two separate processor modules. One processor module consists of two processors each referred to as a switch processor (SP), while the other module has a single processor called the route processor (RP). The RP module supports the main system processor (which includes a network management processor that runs the system management software), and also a larger portion of the system memory components.

The RP is the processor responsible for executing the system management functions that configure and control the switch/router. In addition to having a high-power microprocessor, the RP supports the following features:

- Three main system memory components:
    - Two DRAM SIMMs that maintain the queues used for storing incoming and outgoing packets. The DRAMs also hold caches required for the system.
    - One Flash memory SIMM (EPROMs) for storing the Cisco IOS software image. The default memory is 8 MB and is upgradeable to 16 MB.
    - Two Flash PC card slots for creating additional Flash memory for storing the configuration information and system software.
- Temperature sensor that allows for the monitoring of the internal system environment.
- Console port that can be used to connect a terminal or a modem to the switch/router for system configuration and management.
- 10/100 Mb/s Ethernet port that can be used to connect the switch/router to a management device with an Ethernet interface or an SNMP management station.

In addition to these features, the RP performs the following management functions:

- Monitoring the switch/router interfaces and the environmental status of the whole system.
- Providing SNMP management and the console (Telnet) interface used for system management.

The RP, like the SRP on the Catalyst 8510, runs the Cisco IOS software that implements the unicast and multicast routing protocols and constructs and maintains the distributed forwarding tables used in the line cards. The SNMP agents and the MIBs used for the switch/routers management, as well as the advanced management applications used for traffic management, run in the RP.

The SP, on the other hand, runs the Layer 2 control plane protocols such as Spanning Tree Protocol (STP), IEEE 802.1AB Link Layer Discovery Protocol (LLDP), and VLAN Trunking Protocol (VTP). Together, the combined functions of the two SPs and the RP in the Catalyst 8540 are logically equivalent to the functions of the SRP in the Catalyst 8510.

The two Catalyst 8540 SPs take up two slots in the 13-slot chassis, with a third slot reserved for a redundant SP. If any one of the two SPs fails, the third redundant SP will take over. One slot in the chassis is reserved for the RP, which is responsible for running system management and control plane software. A second slot is reserved for a redundant RP. With five slots taken up by the SPs and RP and their redundant processors, the remaining eight slots in the Catalyst 8540 are used for line card modules.

10.3   THE SWITCH-ROUTE AND ROUTE PROCESSORS

The Catalyst 8540 employs a 40 Gb/s, shared memory switch fabric, while the Catalyst 8510 employs a 10 Gb/s shared memory fabric. These shared memory switch fabrics allow for full nonblocking transfer of data between ports and system modules. The switch/router ports include 10/100 Mb/s Ethernet, Gigabit Ethernet, and 155 Mb/s/622 Mb/s ATM ports.

The Catalyst 8500 CSR has a distributed forwarding architecture where all line cards in the system can forward packets locally. The system processor (SRP or RP) ensures that the Layer 2 and Layer 3 forwarding information in the line cards are up-to-date. The forwarding tables in the line cards are updated and kept synchronized with the master forwarding table (in the SRP or RP) whenever routing and network topology changes occur.

The system processor (i.e., the SRP in the Catalyst 8510 and RP and SPs in the Catalyst 8540) is the entity responsible for managing almost all aspects of the system operation. It is responsible for running all the routing protocols, constructing and maintaining the routing tables from which the forwarding table (also called forwarding information base (FIB)) is generated and distributed to the line cards. The SRP and the SP are responsible for Layer 2 MAC address learning and distribution to the line cards. The SRP and RP (in the Catalyst 8540) are also responsible for system management and configuration.

Figure 10.2 shows a high-level view of the Catalyst 8500 switch/router architecture. It should be noted that, despite the shared memory bandwidth and system processor differences between the Catalyst 8510 and Catalyst 8540, they have identical functions. The SP and RP in the Catalyst 8540 can be viewed as a single logical processor with functions similar to the SRP in the Catalyst 8510.

The system processor is responsible for all Layer 2 address learning, Layer 3 route determination, and distribution to the line cards. Given that the Catalyst 8500

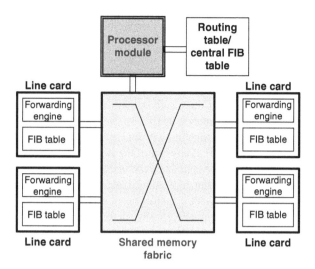

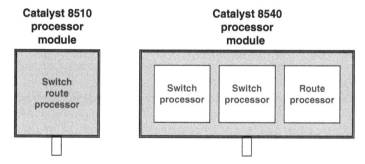

FIGURE 10.2   High-level architecture of the Catalyst 8500 CSR.

is designed as a distributed forwarding system, the system processor must ensure that all Layer 2 and Layer 3 addresses and routes are maintained and updated in the line cards as network changes occur. The system processor is also responsible for all system management, including SNMP, remote monitoring (RMON), and statistics collection.

The system processor runs a number of protocols such as Hot Standby Router Protocol (HSRP), which is a Cisco proprietary protocol (whose standards-based alternative is Virtual Router Redundancy Protocol (VRRP)) for establishing a redundancy fault-tolerant default gateway for a network, Protocol Independent Multicast (PIM) for multicast routing, and a full set of routing protocols, including Routing Information Protocol (RIP), RIP version 2, Open Shortest Path First (OSPF), Interior Gateway Routing Protocol (IGRP), enhanced IGRP, and Border Gateway Protocol -4 (BGP-4).

10.4   SWITCH FABRIC

The Catalyst 8510 employs a 3 MB shared memory with 10 Gb/s of total system bandwidth. The Catalyst 8540 uses a 12 MB shared memory with 40 Gb/s of total system bandwidth. Each architecture is completely nonblocking, allowing all input ports to have equal and full access into the shared memory for packet storage and forwarding. The shared memory is also dynamic, allowing packets stored in memory to consume as much memory dynamically as they need.

Access to the shared memory (for writes and reads) is dynamically controlled by a direct memory access (DMA) ASIC. Given that the shared memory switch fabric is nonblocking, the switch/router does not have to use per-port buffering. This is because the shared memory fabric bandwidth is greater than the combined bandwidth of all the system ports.

Since the architecture is logically an output buffered one, congestion only occurs at an individual output port when the port's resources are oversubscribed. The Catalyst 8500 also supports four priority queues per port and an output port Frame Scheduler that services the output queues based on the priority of each queue.

Each of the line card modules fits into a chassis slot and connects to the Catalyst 8500 shared memory switch fabric. In the Catalyst 8510, each line card is allocated 2.5 Gb/s of the shared memory bandwidth (as shown in Figure 10.3). This allocated bandwidth allows for nonblocking data transfer from any port since each slot is given bandwidth larger than the sum of the bandwidth of all of the ports on the line card.

The 2.5 Gb/s bandwidth allocated to a slot is divided into two: with 1.25 Gb/s allocated to the transmit path and 1.25 Gb/s to the receive path. This ensures that writes and reads to the shared memory can be done in a nonblocking manner, independently, and simultaneously. In the Catalyst 8540, each slot is allocated 5 Gb/s into the shared memory fabric (Figure 10.4). This bandwidth is also divided into 2.5 Gb/s for the transmit and 2.5 Gb/s for receive path to the shared memory.

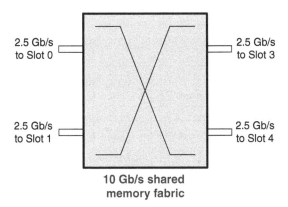

FIGURE 10.3   Switching bandwidth per slot on Catalyst 8510.

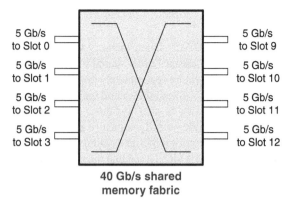

FIGURE 10.4   Switching bandwidth per slot on Catalyst 8540.

Each packet written into the shared memory has an internal routing tag prepended to it. This internal routing tag provides the shared memory switch fabric with the right information to enable it to internally route the packet to the correct destination port(s). The routing tag carries information about the destination port(s), the packet's destination port QoS priority queue, and the packet's drop priority.

A Fabric Interface ASIC (see Figure 10.5) writes the arriving packet into memory after a forwarding decision has been made. The Interface ASIC creates

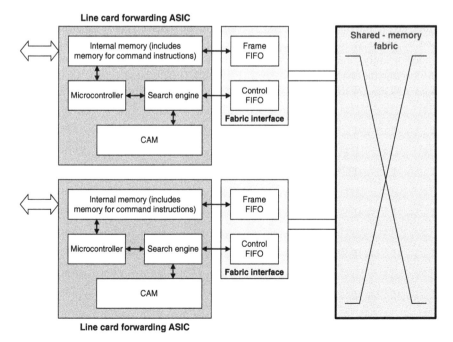

FIGURE 10.5   Catalyst 8500 line card architecture.

a pointer indicating the memory location the packet is stored and generates the internal routing tag that is prepended to the packet and carries the appropriate destination port(s).

The output port Frame Scheduler examines the output port priority queues and then schedules the queued packets out of shared memory using a strict priority, weighted fair queuing (WFQ), weighted round-robin (WRR), or any configured scheduling mechanism (see discussion in the following section).

10.5 LINE CARDS

Each line card contains forwarding ASICs designed to provide interfaces to the shared memory switch fabric, as well as, maintain Layer 2 and Layer 3 forwarding tables. These forwarding tables allow the Catalyst 8500 system to make forwarding decisions at wire speeds before a packet is written into the shared memory.

The system processor is responsible for ensuring that the forwarding tables in the line cards are up-to-date whenever network and routing changes occur. The line cards are also responsible for preparing arriving packets for efficient storage (e.g., segmentation, tagging) in the shared memory switch fabric, QoS policy enforcement, and packet forwarding to the external network.

Figure 10.5 shows the architecture of the Catalyst 8500 line cards. The Catalyst 8500 uses a distributed forwarding architecture where the line cards are equipped with the right forwarding information to make both Layer 3 and Layer 2 forwarding decisions locally at wire speed, as well as enforce QoS and security filtering policies. The distributed forwarding engine ASIC in the line card is responsible for the Layer 2 and 3 address lookups in the CAM table (Figure 10.5), and forwarding of the packet along with its correct Layer 2 address rewrite information to the Fabric Interface. The Fabric Interface (also implemented on the line card) and is responsible for rewriting the Layer 2 addresses in the departing Layer 2 frame carrying the processed packet, QoS classification, and presentation of QoS priority queuing information to the Frame Scheduler.

Each distributed forwarding engine ASIC is assigned four ports on the line card to service. This means two forwarding engine ASICs are required per line card to service eight ports. On the Catalyst 8540, four forwarding engine ASICs are required to service 16 ports. The forwarding engine ASIC also handles all MAC layer functions. The MAC in the 10/100 Mb/s Ethernet ports can run in either full or half duplex mode and is auto-sensing and auto-negotiating, if configured. The distributed forwarding engine ASIC has several key components (Figure 10.5) that are discussed in detail in the following sections.

10.5.1 Internal Memory

Packets arriving at a switch/router port are handled by the Ethernet MAC functions, and then stored in an Internal Memory (Figure 10.5) in the distributed forwarding engine ASIC. This memory consists of a block of SRAM and is 8 kB in total size out of which 2 kB is reserved and used for command instructions. The remaining 4 kB memory is used to store the arriving packet while it waits for the necessary Layer 2 or 3 forwarding table lookup operations to take place.

10.5.2 Microcontroller

The microcontroller in the forwarding engine ASIC is a small processor (mini-CPU) that is used locally to process packets coming from four ports on the line module. The microcontroller supports mechanisms that will allow it to process the arriving packets from the four ports in a fair manner. The scheduling mechanism responsible for the arriving packets ensures that they all have equal access to the Internal Memory.

The forwarding engine ASIC also ensures that forwarding table lookups via the Search Engine are done in a fair manner when the four ports arbitrate for lookup services. Access to the Search Engine is done in a round-robin manner, controlled by the microcontroller that cycles through each port, processing lookup requests as they are submitted.

The microprocessor is also responsible for forwarding special system packets such as routing protocol messages, Address Resolution Protocol (ARP) frames, Spanning Tree BPDUs, Cisco Discovery Protocol (CDP) packets, and other control-type packets to the system processor (SRP, SP, and RP). These special and exception packets are forwarded by the forwarding engine ASIC directly to the system processor.

CDP is a proprietary Cisco Layer 2 protocol that allows Cisco devices to exchange information on the MAC addresses, IP addresses, and outgoing interfaces they support. This feature allows a network manager to have a view of the devices in a network and the interfaces and addresses they support, and also troubleshoot potential network problems.

10.5.3 CAM and Search Engine

The forwarding engine ASIC's Search Engine is responsible for performing the forwarding table lookups to determine correct output port(s) to which a packet should be forwarded. The forwarding tables used by the Search Engine for lookups are stored in a content-addressable memory (CAM), which can store either 16,000 or 64,000 entries as explained earlier.

The Search Engine handles both Layer 2 and Layer 3 forwarding table lookups. It is responsible for maintaining the Layer 2 and Layer 3 forwarding tables (which are in turn generated by the system processor (SRP, SP, and RP). Also, using a hardware-based Access Control List (ACL) feature card (to be discussed later), the Search Engine is capable of performing Layer 4-based lookups based on some Layer 4 fields/information in the packet.

When an arriving packet is being written into the Internal Memory and as soon as the first 64 bytes of the packet are written into the memory, the microcontroller passes to the Search Engine the relevant source and destination MAC address and destination IP address (plus if necessary, the packets Layer 4 information). The Search Engine then uses these extracted packet parameters to perform a lookup in the forwarding tables in the CAM for the corresponding forwarding instructions. The Search Engine uses a binary tree lookup algorithm to locate the output port corresponding to the packets destination MAC address (for Layer 2 forwarding) or the longest network prefix that matches the destination IP address. The Search Engine also retrieves the corresponding MAC address rewrite and QoS information (which is also maintained in the CAM), and forwards this to the control FIFO (Figure 10.5) of the Fabric Interface.

The CAM is designed to have two storage options: one supporting 16,000 entries and the other 64,000 entries. After using the binary tree algorithm to perform the lookup to locate the correct forwarding information, the Search Engine sends the relevant forwarding and MAC address rewrite information to the Fabric Interface for delivery to other processing components. The Layer 2 or 3 lookup provides the forwarding engine ASIC with the destination port for the packet. The packet is then transferred across the shared memory switch fabric to the destination port.

10.5.4 Fabric Interface

After the forwarding table lookup, the Fabric Interface prepares the packet to be sent across the shared memory switch fabric to the destination port. The Fabric Interface has two main components: a control FIFO and a frame FIFO. The Internal Memory of the forwarding engine ASIC is directly connected to the frame FIFO (Figure 10.5), while the Search Engine is directly connected to the control FIFO.

As soon as the Search Engine performs the lookup in the forwarding tables (stored in the CAM), the packet is transferred from Internal Memory to the frame FIFO. In parallel, the MAC address rewrite and QoS information for the packet are forwarded by the Search Engine to the control FIFO.

The Fabric Interface then decrements the IP TTL, recomputes the IP checksum, rewrites the source and destination MAC address in the frame carrying the packet, and recomputes the Ethernet frame checksum. The Fabric Interface prepends an internal routing tag to the packet. This routing tag contains the destination port, the QoS priority queue, and packet discard priority to the packet.

Once completed, the Frame Interface signals the Frame Scheduler to write the packet into the shared memory. At the destination port, the Fabric Interface transfers the packet to its output Ethernet MAC entity for transmission to the external network. Since all relevant processing (IP TTL decrement, IP checksum calculation, MAC address rewrite, Ethernet checksum) has already been performed at the ingress port, no additional processing is required on the packet.

10.6   CATALYST 8500 FORWARDING TECHNOLOGY AND OPERATIONS

The Catalyst 8500 supports a distributed forwarding architecture that uses a forwarding approach where the forwarding information generated by the central system processor is distributed to the individual line card modules to enable them forward packets locally. Distributed forwarding in the line cards results in very high-speed forwarding table lookups and forwarding. This approach provides higher forwarding performance and scalability that is suitable for service provider networks and large campus and enterprise core networks.

10.6.1   Forwarding Philosophy

Some Layer 3 forwarding methods are based on a route/flow cache model where a fast lookup cache is maintained for destination network addresses as new flows go through the system. The route/flow cache entries are traffic driven, in that the first packet of a new flow (to a new destination) is Layer 3 forwarded by the system processor via software-based lookup, and as part of that forwarding operation, an entry for that destination is added to the route/flow cache.

This process allows subsequent packets of the same flow to be forwarded via the more efficient route/flow cache lookup. The route/flow cache entries are periodically aged out to keep the route/flow cache fresh and current. The cache entries can also be immediately invalidated if network topology changes occur.

This traffic-driven approach of maintaining a very fast lookup cache of the most recent destination address information is optimal for scenarios where the majority of traffic flows are long flows. However, given that traffic patterns at the core of the Internet (and in some large campus and enterprise networks) do not follow this "long flow" model, a new forwarding paradigm is required that would eliminate the increasing cache maintenance resulting from the growing numbers of short flows and dynamic network traffic changes.

The distributed forwarding architecture avoids the high overhead of continuous route/flow cache maintenance by using a full forwarding table for the forwarding decisions. The forwarding table contains the same forwarding information in the main routing table maintained by the system processor. The maintenance of the same critical information required for forwarding packets in the forwarding table and routing table eliminates the need for maintaining a route/flow cache for packet forwarding.

The distributed forwarding architecture best handles the network dynamics and changing traffic patterns resulting from the large numbers of short flows typically associated with interactive multimedia sessions, short transactions, and Web-based applications. Distributed forwarding offers high forwarding speeds, a high level of forwarding consistency, scalability, and stability in large dynamic networks.

Additionally, distributed forwarding in the line cards is less processor intensive than forwarding by the main system processor (CPU) because the forwarding decisions made by each line card are on a smaller subset of the overall destination addresses maintained in the system. Basically, in distributed forwarding, the system processor (CPU) offloads a majority of its processing to the line cards, resulting in efficient and higher overall system performance. This architecture allows for high-speed forwarding (wire speed on all ports) with low data transfer latency. One other key benefit of distributed forwarding is rapid routing convergence in a network.

Since the forwarding table is distributed to all line cards, any time a route is added, a route flaps, or a route goes away, the system processor updates its forwarding table with that information and also updates the forwarding tables in the line cards. This means that system processor interrupts are minimized, because there is no route/flow cache to invalidate and flow destinations to relearn for the cache. The line cards are able to receive the new routing updates quickly (via the system processor forwarding table) and the network reconverges quickly around a failed link base if that happens.

When designing a distributed forwarding system (e.g., the Catalyst 8500), it is highly beneficial to separate the control plane and data forwarding plane functions in the system. This design approach allows the system processor to handle the control plane processing while the line cards handle (without system processor intervention) the data forwarding. Other than interacting with the system processor to maintain their local forwarding tables, the line cards operate almost autonomously.

The system processor handles all routing and system-level management functions, such as running the unicast and multicast routing protocols and constructing and maintaining the routing table and distributed forwarding tables (used by the line cards), ARP, CDP, and STP configuration. Each distributed forwarding engine ASIC (in the Catalyst 8500) is responsible for identifying these special control and management packets and forwarding them to the system processor.

The system processor is responsible for running all of the Catalyst 8500's routing protocols that are implemented as part of the Cisco IOS software. Most importantly, the system processor is responsible for maintaining the routing and forwarding tables. Using these capabilities, the system processor creates a master forwarding table, which is a subset of the information in the routing table and contains the most important information required for routing packets.

The forwarding table reflects the true current topology of a network, thus allowing high-speed data forwarding to be done using the actual network topology information. The master forwarding table is then copied to the line cards, allowing them to make Layer 3 forwarding decisions locally without system processor intervention. This forwarding architecture allows the Catalyst 8500 to forward packets at wire speed at all ports.

The system processor is also responsible for multicast routing and forwarding where it maintains state information regarding multicast sessions and group memberships. The Catalyst 8500 supports the older Distance Vector Multicast Routing Protocol (DVMRP), as well as the PIM (sparse mode and dense mode). The system processor also responds to and forwards multicast joins, leaves, and prune messages. However, the line cards are responsible for multicast forwarding using the multicast forwarding tables created by the system processor.

In Layer 2 forwarding, the forwarding decisions are also made at the line cards using the Layer 2 forwarding tables created by the system processor (SRP in Catalyst 8510 or SP in Catalyst 8540). The SRP or SP is responsible for generating the Layer 2 forwarding information as well as running STP and handling bridge group configuration. The system processor is responsible for running all STP functions that include determining the spanning tree root bridge, optimum path to the root, and forwarding and blocking spanning tree links.

10.6.2 Forwarding Operations

The Catalyst 8500 distributed forwarding architecture uses two types of tables – a Layer 3 forwarding table (FIB) and an adjacency table (Figure 10.6) – that are maintained by the system processor (i.e., the SRP in the case of the Catalyst 8510, and RP and SP in the Catalyst 8540). These tables are downloaded to the line cards for local use. The Layer 3 forwarding table is a mirror of the main forwarding information in the routing table and is used for making Layer 3 forwarding decisions.

The adjacency table maintains information about the nodes that are directly connected (adjacent) to the switch/router (physically or logically through a virtual connection). This table contains the Layer 2 information about the adjacencies such as the Layer 2 (MAC) addresses of their connected interfaces. The Layer 2 addresses are used during the packet rewrite process when packets are prepared to be sent to the next hop (which is an adjacent node). Each entry in the Layer 3 forwarding table (i.e., the next hop IP address that matches a destination address in a packet) includes a pointer to a corresponding adjacency table entry.

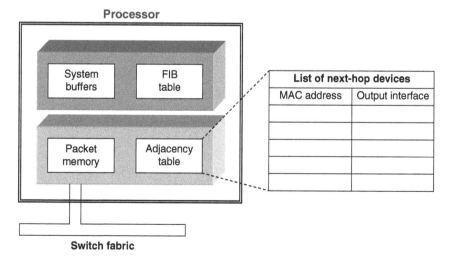

FIGURE 10.6 Forwarding information base (FIB) and adjacency table.

The main Layer 3 forwarding table maintained by the SRP or RP is constantly updated by inputs from the routing table which in turn is maintained by the routing protocols. After the SRP or RP resolves a route, the resolution translates to a next hop IP node, which is an adjacent node. The SRP or RP then copies this updated information to the line cards, allowing them to maintain an updated and current view of the network topology.

These routing updates enable correct forwarding decisions to be made, as well as fast network convergence when routing topology changes take place. As the network topology changes dynamically, the routing table is updated and the changes populated in the Layer 3 forwarding table. The SRP or RP modifies the Layer 3 forwarding table any time a network route is removed, added, or changed in the routing table. These updates are immediately copied to the forwarding tables in the line cards. This results in all line cards having a correct view of the network topology at all times.

Adjacency table entries are added when a routing protocol detects that there is a next hop (i.e., an adjacent node) leading to an IP destination address. The adjacency table is updated when a network route is removed, added, or changed in the routing table. The updates in the adjacency table can be from inputs from the routing protocols, which include adjacencies derived from next hop IP node information and multicast (S,G) interfaces carrying multicast traffic flows to multicast groups.

When a packet arrives at a Catalyst 8500 switch/router port, the distributed forwarding engine ASIC performs a lookup in its forwarding table using the packet's destination IP address. The longest matching forwarding table entry produces a next hop IP address that also points to a corresponding adjacency entry. This adjacency table entry provides the Layer 2 address of the receiving interface of the next hop, which can be used to rewrite the Layer 2 destination address of the outgoing Layer 2 frame at the outgoing interface. The Layer 2 frame carrying the outgoing packet is forwarded to the next hop using the discovered adjacent table Layer 2 information.

10.6.2.1 Layer 3 Forwarding in the Catalyst 8500 The packet forwarding process in the Catalyst 8500 can be described by the following steps:

1. When a packet is received at a switch/router port, the packet is first handled by the MAC sublayer functions in the distributed forwarding engine ASIC. After this, the packet is stored in the forwarding engine ASIC's Internal Memory.

2. As soon as the distributed forwarding engine ASIC receives the first 64 bytes of the arriving packet, the microcontroller's microcode reads the packet's source and destination IP addresses. If the packet's destination MAC address is the MAC address of switch/router's receiving interface, then packet requires Layer 3 forwarding. If not, it requires Layer 2 forwarding.

3. The MAC and IP destination addresses of the packet are used by the forwarding engine ASIC's Search Engine to perform a lookup in the CAM for the best matching entry.

4. After the best matching entry is discovered, the result is sent back to the microcontroller.

5. The microcontroller then moves the packet from the forwarding engine ASIC's Internal Memory to the frame FIFO in the Fabric Interface. In parallel, the Search Engine sends all relevant QoS classifications, priority queuing, and MAC address rewrite information to the Control FIFO also in the Fabric Interface.

6. The input Fabric Interface then carries out all the packet rewrite (IP TTL decrement, IP checksum computation, source and destination MAC address, Ethernet checksum computation) and QoS classifications.

7. The input Fabric Interface prepends an internal routing tag to the packet. The internal routing tag identifies the QoS priority queuing for the packet, the destination port(s), and the packet's discard priority. The QoS priority indicated in the tag determines which of the four priority queues maintained at the destination port the packet will be queued.

8. As soon as the packet is completely transferred to the Fabric Interface's Frame FIFO, the Frame scheduler is signaled (based on the packet's QoS priority level) to initiate arbitration for access to the shared memory. When access is granted, the packet is transferred into the shared memory. The packet is stored along with a pointer that indicates the destination port of the packet.

9. The Fabric Interface signals the destination port to retrieve the packet out of a known location in the shared memory. The internal routing tag indicates to the destination port that it is receiving the correct packet.

10. The destination port then forwards the packet to the external network.

In the Catalyst 8500, the multicast routing table is also centralized and maintained on the SRP (in Catalyst 8510) or RP (in Catalyst 8540). The forwarding engine ASIC in a line card consults the forwarding table (that includes multicast routes) to forward the multicast packets to appropriate destinations.

A multicast routing table is not the same as a unicast routing table. A multicast routing table maps a source IP address transmitting multicast traffic to a multicast group of receivers interested in the multicast traffic (the mapping commonly represented by (S,G)). The table consists of an input interface (port) on the switch/router and a set of corresponding output interfaces (ports).

The central multicast routing table maintained by the multicast routing protocols running in the SRP or RP is distilled into multicast forwarding information to be used in the line cards. By distributing forwarding engines and the associated distribution of the multicast forwarding information in the Catalyst 8500, the line cards are able to forward multicast traffic locally based on the multicast topology contained in distributed information.

The local multicast forwarding information allows an input port on a line card to determine which output interfaces on the switch/router require a particular multicast traffic. The input line card then signals the shared memory switch fabric about which output ports to forward that traffic to. Changes in the multicast routing table are instantly distilled and copied to the line cards, allowing them to maintain up-to-date multicast distribution map of the multicast traffic in a network.

10.6.2.2 Layer 2 Forwarding in the Catalyst 8500 As already stated, if the destination MAC address of an arriving frame is the switch/router's interface MAC address, then the packet is Layer 3 forwarded – if not, it is Layer 2 forwarded. When groups of ports or a port in the switch/router are configured to run in Layer 2 forwarding (bridging) mode, the forwarding engine ASIC's Search Engine performs a lookup in the CAM based on the Layer 2 MAC address of the arriving packet.

Given that the Catalyst 8500 has a distributed forwarding architecture, each distributed forwarding engine ASIC maintains a list of MAC addresses and their corresponding exit ports that are of local significance. For example, if Address 01:23:45:67:89:ab is a destination MAC address learned on switch/router port SREthernet 0/2, the remaining ports on the Catalyst 8500 do not have to store this MAC address in their Layer 2 forwarding tables (in the CAMs) unless they have a packet to forward to Address 01:23:45:67:89:ab (at port SREthernet 0/2).

The system processor (SRP or RP) has a central CAM (that maintains a master (integrated) forwarding table that holds both Layer 2 and Layer 3 addresses. When the distributed forwarding ASIC learns a new MAC address, that MAC address (not the packet that carries it) is forwarded to the system processor so that it has an updated list of all MAC addresses learned. The system processor populates its CAM with the new MAC address learned. The system processor's central CAM contains all MAC addresses manually configured in the CAM plus those that the switch/router has learned dynamically.

If an arriving Ethernet packet has destination MAC address that is ffff.ffff.ffff (i.e., the broadcast address), then the packet is prepended with an internal routing tag that indicates that its destination is all ports in the bridge group the port belongs to. The tagged packet is then sent to the shared memory switch fabric for storage. The Fabric Interface creates a pointer to the packet's memory location that is signaled to all the ports in that bridge group. This means that if six ports are configured in a bridge group, then all six ports would receive the broadcasted packet.

To describe the Layer 2 forwarding process in the Catalyst 8500, let us assume that the source and the destination MAC address of a packet have already been learned. The following steps describe the Layer 2 forwarding process:

1. When a packet arrives at a switch/router port, the MAC-layer functions in the distributed forwarding engine ASIC process the packet, and the packet is placed in the forwarding engine ASIC's Internal Memory.

2. As soon as the distributed forwarding engine ASIC receives the first 64 bytes of the packet, the microcontroller's microcode reads the packet's source and destination MAC addresses. If the packet's destination MAC address is not that of the receiving port of the switch/router, then Layer 2 forwarding is required. The MAC address information is then passed to the Search Engine for the Layer 2 address lookup in the CAM.

3. Let us assume the packet has been transmitted from a station in a particular VLAN (i.e., a Layer 2 broadcast domain). The Search Engine performs a lookup in the CAM for the destination port entry that corresponds to the packet's destination MAC address.

4. When the correct destination port is located, the microcontroller transfers the packet from the distributed forwarding engine ASIC's Internal Memory to the Frame FIFO in the Fabric Interface. At the same time, the Search Engine forwards the packet's QoS classification and priority queuing information to the Fabric Interface's Control FIFO. An internal routing tag that is prepended to the packet identifies the destination port and the QoS priority queuing for the packet.

5. As soon as the packet is completely transferred into the Frame FIFO, the Frame Scheduler is signaled to start arbitrating for access to the shared memory. When the Frame Scheduler is granted access, it transfers the packet into the shared memory.

6. The Fabric Interface then signals the destination port ASIC to read the packet from the shared memory. The internal routing tag indicates to the destination port that it is receiving the correct packet.

7. The destination port then transmits the packet out to the external network.

A bridge group refers to a (Layer 2) broadcast domain configured within the switch/router. Typically, to simplify network design (although not necessarily), a bridge group is configured to correspond to a particular IP subnet. Up to 64 bridge groups can be supported in the Catalyst 8500.

It is important to note that a bridge group is different from a VLAN. A VLAN is a broadcast domain that terminates at router (Layer 3 forwarding device). Inter-VLAN communication can only take place through a router because this is where the Layer 3 forwarding from one VLAN to the other is expected to take place. An IEEE 802.1Q VLAN trunking standard can trunk multiple VLANs, each VLAN terminating at a router port.

If a VLAN needs to be extended through a switch/router, then the extension can be done by configuring a bridge group. In this case the switch/router is transparent to a particular VLAN traffic going through it. The VLAN on both sides of a bridge group are essentially the same one VLAN (same broadcast domain). On the other hand, two VLANs configured on a network and communicating via Layer 3 forwarding in the switch/router is not a bridge group since that traffic is not bridged through the switch/router.

10.7   CATALYST 8500 QUALITY-OF-SERVICE MECHANISMS

QoS has become increasingly more important as networks carry delay sensitive end-user traffic such as streaming video, voice, and interactive applications that require low latencies. The Catalyst 8500 supports a number of QoS mechanisms

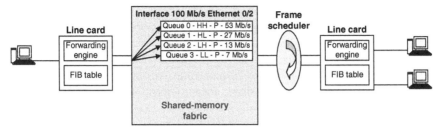

FIGURE 10.7 Traffic scheduling and bandwidth allocation.

that are incorporated into the switch/router architecture. The Fabric ASIC and Frame Schedule are the two main QoS components in the Catalyst 8500. They perform the packet classification and priority queuing at the input port, in addition to packet scheduling of the priority queues using a WRR algorithm at the output port.

The Catalyst 8500 series supports four priority queues per port in which packets can be queued based on, for example, their IP Precedence bit settings (Figure 10.7). The IP Precedence information can be extracted from the IP header type-of service (ToS) field (also called the Service Type field) or the corresponding DSCP bits in the IP header. In the IP packet, the first 3 bits of the ToS field are used to signal the delay and drop priority of the packet.

The rightmost bit in the 3 bits used for the IP Precedence defines the drop priority of a packet. If this bit is set (to 1), the Catalyst 8500 drops that packet when the destination queue becomes full before it drops packets with bits not set. The leftmost 2 bits are used to define the delay priority of a packet. The eight priority classes resulting from the 3 bit IP Precedence translate to eight different classes in the Catalyst 8500. The eight traffic classes are mapped to the four QoS priority queues in the Catalyst 8500 as summarized in Table 10.1.

All control and management traffic, such as routing protocol updates, STP BDPU information, and management packets (ARP, IGMP, ICMP, etc.), are placed in the highest-priority queue for transmission to the system processor (SRP in Catalyst 8510 or RP in Catalyst 8540). These special packets have to be forwarded to the system processor with minimum delay at the input port.

TABLE 10.1   IP Precedence Values

IP ToS Field Value	Delay Priority	Drop Priority	Queue Selected
0 0 0	0 0	0 (Drop packet first)	QoS-3 (lowest priority)
0 0 1	0 0	1 (Drop after last)	QoS-3
0 1 0	0 1	0	QoS-2
0 1 1	0 1	1	QoS-2
1 0 0	1 0	0	QoS-1
1 0 1	1 1	1	QoS-1
1 1 0	1 1	0	QoS-0
1 1 1	1 1	1	QoS-0 (highest priority

The Catalyst 8500 queue packets are based on two parameters: the delay priority setting in the IP Precedence bits in the packet and the target next hop interface. The delay priority setting (i.e., the leftmost 2 bits) of the IP Precedence bits signal to the Frame Scheduler which of the four priority queues the packets should be queued in. The Fabric Interface then supplies a pointer to the output (destination) port, indicating the priority queue and shared memory location from which to extract (read) the packet.

Recall that the five-slot modular Catalyst 8510 chassis can support up to 32 10/100 Mb/s Ethernet ports or 4 GbE ports. This means the Catalyst 8510 can support a maximum of 32 next hop interfaces. Each of these 32 possible next hop ports supports four priority queues. Packets processed by any distributed forwarding engine ASIC can be queued in 1 of 128 queues (equal to 32 ports × 4 queues) in the shared memory based on delay priority settings in the packet and next hop interface the packet is to be sent to. The Fabric Interface provides the output (destination) port with a pointer to the priority queue and shared memory location of the packet.

Each priority queue at the Catalyst 8500 port is configured with a higher queue threshold and a lower queue threshold limit. These queue limits are user config-urable to meet targeted traffic management policies. These two queue limits can be viewed, respectively, as the in-of-profile queue threshold (for packets conforming to the configured traffic policy) and out-of-profile queue threshold.

A queue accepts all packets if its current queue length is below the configured lower queue threshold. If the queue length is between the lower and higher queue threshold limits, the queue accepts only packets with drop priority setting of 0 (or in-profile packet) and discards packets with drop priority setting of 1 (out-of-profile). When queue length is greater than the higher queue threshold, then all arriving packets are discarded until the congestion subsides.

The Frame Scheduler performs two main functions in the Catalyst 8500. It is responsible for scheduling packets that have been processed by the distributed forwarding engine ASIC into the shared memory switch fabric based on the priority queuing requirement of the packet. The Frame Scheduler also schedules packets out of the shared memory switch fabric at the output ports based on the WRR algorithm.

To write packets into the shared memory, the distributed forwarding engine ASIC sends a request to the Frame Scheduler requesting access to the shared memory. The Frame Scheduler receives requests from all active forwarding engine ASICs and processes them in a time-division multiplexing (TDM) fashion. With this mechanism, each forwarding engine ASIC is always given an equal opportunity to write a complete packet into the shared memory when it is granted access.

Recall also that each distributed forwarding engine ASIC on a line card is responsible for four ports. This means that the Frame Scheduler allows the distributed forwarding engine ASIC to write a maximum of four packets into the shared memory. Each packet written into the shared memory is assigned an internal routing tag, which (as mentioned earlier) contains the destination port, priority queuing, and packet discard priority of the packet. Using the internal routing tag, the input Frame Scheduler stored the packet in the shared memory in the correct priority queue (see Figure 10.7).

The "LL," "LH," "HL," and "HH" designations in Figure 10.7 refer to the grouping of the IP Precedence bit settings used by the Catalyst 8500 to map IP Precedence-marked traffic to the appropriate priority queue. The Catalyst 8500 maintains (not shown Figure 10.7) a fifth high-priority internal routing tag that is prepended to all control and management packets. This fifth routing tag signals that such critical packets require immediate delivery to the system processor.

At the output side of the shared memory switch fabric, the Frame Scheduler is responsible for scheduling packets from each priority queue using the WRR algorithm. The WRR algorithm allows a network manager to configure scheduling weights that define how much bandwidth each queue should receive. Under conditions where there is no congestion, the WRR mechanism and the scheduling weights configured do not really affect how packets are transmitted out of the shared memory switch fabric, because there is an abundance of bandwidth out of the queues. However, if the output is congested due to excess input traffic, then the WRR mechanism schedules each queue at a port based on the priority setting defined by the weights.

The Catalyst 8500 also allows a network manager to override the global QoS settings (configured through the switch/router management interface in the system processor) by defining priority configuration on a per port basis. The network manager also has the option of configuring bandwidth and packet classification (and queuing) based on source–destination address pair, destination address, or source address basis. The manager can configure scheduling weights to allow certain IP addresses to have more bandwidth than others.

11

QUALITY OF SERVICE MECHANISMS IN THE SWITCH/ROUTERS

11.1 INTRODUCTION

The QoS mechanisms discussed in this chapter are based on the Cisco Catalyst 6500 Series of switch/routers. The Catalyst 6500 family of switch/routers supports a wide range of QoS features, which makes this discussion representative of the main QoS features found in the typical switch/router in the market. This case study allows the reader to appreciate better the kind of QoS features the typical switch/router would support.

The ever-growing demands for higher network performance and bandwidth, and the greater need for smarter network applications and services, are some of the factors driving the design of today's switches, routers, switch/routers, and other network devices. Users continue to demand a wide range of performance requirements and services that include high QoS, network and service availability, and security.

These demands also drive the need to build scalable and more reliable enterprise and service provider networks. QoS mechanisms, in general, represent a collection of capabilities that provide ways to identify, prioritize, and service different classes of traffic in a network. This identification and classification allows the network to prioritize and service traffic such that end-user requirements are satisfied as best as possible.

Switch/Router Architectures: Shared-Bus and Shared-Memory Based Systems, First Edition. James Aweya.
© 2018 The Institute of Electrical and Electronics Engineers, Inc. Published 2018 by John Wiley & Sons, Inc.

The Catalyst 6000/6500 Series support a number of QoS tools that are used to provide preferential treatment of traffic as they pass through the switch/router. Some Cisco Catalyst 6500 switch/routers perform QoS processing centrally on the Supervisor Engine, while those with distributed processing architectures perform the processing directly on the line cards. The Catalyst 6500 employs some hardware support to perform QoS processing.

The various Catalyst 6500 switch/router components that are involved in packet forwarding and QoS processing are the Multilayer Switch Feature Card (MSFC), Policy Feature Card (PFC), and the network port Application-Specific Integrated Circuits (ASICs) on the line cards. The PFC is primarily responsible for hardware-based Layer 2 and 3 forwarding of packets, as well as supporting a number of important QoS functions.

The PFC supports the necessary ASICs needed to perform hardware-based Layer 2 and 3 forwarding, QoS classification and priority queuing, and security access control list (ACL) filtering. The PFC, being mainly a forwarding engine, requires a route processor (i.e., the MSFC) to populate a Layer 3 route/flow cache or full forwarding table used by its Layer 3 forwarding engine ASIC.

If no route processor is installed (as is possible in some earlier Catalyst 6500 switch/router configurations using Supervisor Engine 1A), the PFC can perform only Layer 3/4 QoS classification and security ACL filtering but not Layer 3 packet forwarding. The reason for this limitation is that the MSFC is required to provide the route processor functions needed by the PFC to perform Layer 3 forwarding of packets. The MSFC provides the necessary forwarding information maintained in the PFC's route/flow cache or forwarding table (or forwarding information base (FIB)) without which the PFC cannot Layer 3 forward packets.

In some Cisco Catalyst 6500 platforms, the line card is the main component that performs QoS processing. In these platforms, the QoS features supported are primarily implemented in the network port ASICs to allow for high-speed processing of arriving traffic. The level of QoS processing a line card supports depends on the functionality built into the line card port ASIC.

Over the development cycle of the Catalyst 6500, and with advancements in hardware and software technology, a number of QoS tools have been developed and are available for the Catalyst 6500. For this reason, the QoS processing capabilities differ between the different generations of the Cisco Catalyst 6500 Series. This chapter presents an overview of the main QoS features available on Cisco Catalyst 6000 and 6500 switch/routers (see Chapters 7 and 9).

11.2   QoS FORWARDING OPERATIONS WITHIN A TYPICAL LAYER 2 SWITCH

This section describes the basic operations performed on an Ethernet frame as it passes through an Ethernet (Layer 2) switch. The main components involved in the forwarding operations are illustrated in Figure 11.1. Most of the simpler Layer 2

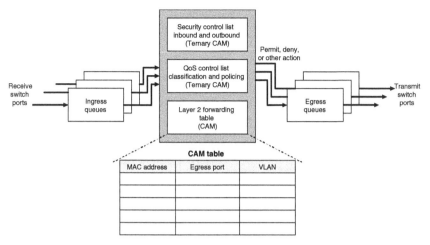

FIGURE 11.1   Main forwarding components within the Layer 2 switch.

switches have only a Content Addressable Memory (CAM) that holds the Layer 2 address table required for Layer 2 forwarding. Some of the higher end Layer 2 switches, on the other hand, support, in addition, Ternary CAMs (TCAMs) that maintain entries used for QoS and security processing (not for Layer 3 forwarding). Layer 3 devices support TCAMs that maintain the Layer 3 routes and addresses that are learned from the routing protocols.

The CAM and TCAM form part of the most important components used in the hardware processing and forwarding of packets, and switches, switch/routers, and routers leverage these for wire speed Layer 2 and 3 forwarding. Typically, these architectures support the ability to perform multiple searches or lookups in different parts of the CAM and TCAM simultaneously. This ability to perform multiple lookups in parallel allows the device to forward packets at higher speeds.

When a Layer 2 packet enters a port on a Layer 2 switch, it is stored in one of the port's ingress (input) queues right after Layer 2 forwarding table lookup (in some architectures), or before forwarding table lookup (in other architectures). The input queues can be configured with QoS priorities with each queue given a different service and packet discard priority profile.

A higher queue service priority ensures that time-sensitive traffic is not held behind nontime-sensitive traffic during network congestion periods. Scheduling algorithms such as strict priority, weighted fair queuing (WFQ), weighted round-robin (WRR), and deficit round-robin (DRR) can be used to service the ingress priority queues.

The process of receiving, classifying, priority queuing, and servicing a packet at the input port also requires the switch to perform a number of other important tasks. These tasks include performing a forwarding table lookup to find the egress switch port(s) to which the packet is to be forwarded, and applying security policies to the packet. These tasks can be executed in a pipeline or in parallel by the forwarding engine using the modules shown in Figure 11.1. These modules are described in Table 11.1.

TABLE 11.1 Forwarding and QoS Modules in the Layer 2 Switch

Module	Description
Layer 2 forwarding table	The Layer 2 forwarding (address) table is typically implemented in a CAM. The packet's destination MAC address is read by the forwarding engine and used as a key (an index) into the CAM holding the MAC address table. If the destination MAC address is found, its corresponding egress port and, if applicable, VLAN ID are read from the CAM. If the MAC address is not found, then the packet is marked for broadcast through all other switch ports (a process called flooding). This involves forwarding the packet out every switch port in the VLAN the packet belongs to
QoS ACLs	The switch can be configured with ACLs to classify incoming packets according to certain parameters in the packet. This allows a network manager to priority queue, police, shape the rate of traffic flows, and to write/remark QoS markings in outbound packets. The QoS ACLs are typically implemented in a TCAM to allow these QoS decisions to be made in a single table lookup
Security ACLs	To implement security policies, the switch can be configured with security ACLs to identify arriving packets according to their MAC addresses, IP addresses, protocol types, and Layer 4 port numbers. The ACLs are typically implemented in a TCAM such that a security filtering decision on whether to forward or filter a packet can be made in a single table lookup

As illustrated in Figure 11.1, after the forwarding table lookups in the CAM and TCAM have been performed, the packet is transferred into the appropriate priority queue at the outbound switch port. The egress priority queue is determined by either the QoS bit settings carried in the packet or in an internal routing tag prepended to the packet. Similar to the ingress priority queues, the egress priority queues can also be serviced according to priority, which can depend on the time criticality of the queued traffic using scheduling algorithms such strict priority, WFQ, WRR, and DRR.

11.3 QoS FORWARDING OPERATIONS WITHIN A TYPICAL MULTILAYER SWITCH

As discussed in Chapter 3, the route/flow cache-based switch/router architectures employ a route processor and a forwarding engine where the route processor determines the destination (next hop node and outbound port) of the first packet in a traffic flow (typically, via software-based Layer 3 forwarding table lookup). The forwarding engine receives the resulting destination information of the first packet from the route processor and sets up a corresponding entry in its route/flow cache.

The forwarding engine then forwards subsequent packets in the same traffic flow based on the newly created flow entry in its cache. Even in architectures that are not route/flow cache based, the same flow caching technique can still be used to generate traffic flow information and statistics.

The topology-based switch/router architectures employ a forwarding engine (which sometimes is a specialized forwarding hardware) and a full Layer 3 forwarding table for packet forwarding. A route processor runs Layer 3 routing protocols to construct a routing table that reflects the true view or topology of the entire network. The contents of the routing table are distilled to generate the Layer 3 forwarding table used by the forwarding engine.

The topology-based architecture allows for efficient forwarding table lookup, typically, in hardware, and can support high packets forwarding rates. For a given destination IP address, the longest matching prefix in the forwarding table provides a corresponding next hop IP address and switch/router port out of which the packet should be forwarded to get to its destination.

As the network topology changes over time, the routing table and forwarding table are updated and packet forwarding continues without noticeable performance penalty. Typically, a route updating process running on the route processor downloads the current routing table information to forwarding table used by the forwarding engine.

To support high-speed forwarding rates, high-end, high-performance switch/ routers forward packets using specialized forwarding engine ASICs. As illustrated in Figure 11.2, specific Layer 2 and Layer 3 components, including forwarding tables and ACLs, are maintained in hardware. Layer 2, Layer 3, ACL, and QoS

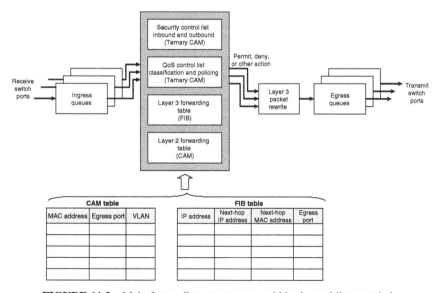

FIGURE 11.2   Main forwarding components within the multilayer switch.

policy tables are stored in high-speed memory so that forwarding decisions and filtering can be done at wire speed.

The switch/router performs lookups in these tables to determine the forwarding instructions for a packet. Along with specifying which port to forward a packet out of, the instructions indicate whether a packet with a specific destination IP address (and other parameters) is supposed to be dropped according to an ACL filtering rule.

As illustrated in Figures 11.1 and 11.2, switches maintain these tables using specialized memory architectures, typically, CAMs and TCAMs. The Layer 2 table is typically maintained in a CAM to allow the Layer 2 forwarding engine to make Layer 2 forwarding decisions at high speeds. TCAMs are mostly used for maintaining Layer 3 forwarding tables and allow for lookups based on longest prefix matching. IP routing and forwarding tables are normally organized by IP address prefixes.

The TCAM can also be used to store QoS, ACL, and other information generally associated with upper-layer protocol information processing in a packet. Most switches, switch/routers, and routers support multiple TCAMs to allow for both inbound and outbound QoS, as well as security ACLs, to be processed in parallel when a Layer 2 or Layer 3 forwarding decision is being made.

The term VMR (Value, Mask, and Result) is often used to describe the format of entries maintained in a TCAM. The "Value" in VMR refers to the pattern stored in the TCAM that can have some of its bits masked (i.e., concealed) by the "Mask." The masked "Value" is to be matched by fields extracted from a packet header fed into the TCAM. Examples of extracted packet fields include IP addresses, protocol ports, DSCP values, and so on.

The "Mask" refers to the bits used to conceal the associated "Value" pattern in the TCAM. The masked "Value" provides/determines the prefix that is to be matched by the extracted packet fields fed into the TCAM. The "Result" refers to the result or action pointed to when a search or lookup in the TCAM hits the masked "Value" pattern.

This "Result" returned could be a "permit" or "deny" action stored in the TCAM for QoS or security ACLs filtering. The "Result" values could also be for priority queuing in a QoS policy in the case of a TCAM used for traffic classification and priority queuing. The "Result" could also be a pointer to an entry in a Layer 2 adjacency table that contains the next hop port and MAC address rewrite information in the case of a TCAM used for IP packet forwarding.

Figure 11.2 shows the main Layer 2 and 3 forwarding and QoS modules in a typical switch/router. Similar to a Layer 2 switch, packets arriving on a switch/router port are stored in the appropriate ingress priority queue right after forwarding table lookup (or just before lookup depending on the forwarding architecture used). In the latter, each packet is placed in an ingress queue and its header is examined to retrieve both the Layer 2 and Layer 3 destination addresses to be used for the forwarding table lookup.

The switch/router has to determine to which switch/router port(s) to forward the packet after the Layer 2 or 3 forwarding table lookup. The forwarding engine also

TABLE 11.2   Forwarding and QoS Modules in the Multilayer Switch

Module	Description
Layer 2 forwarding table	Typically, the Layer 2 forwarding table is maintained in a CAM. The destination MAC address in an arriving packet is extracted and used as an index to the CAM. If the packet requires Layer 3 forwarding, its destination MAC address is that of the receiving interface on the switch/router. In this case, the CAM lookup results are used only to decide that the packet should be Layer 3 forwarded. Packets addressed to the MAC address of any of the switch/router local interfaces should be Layer 3 forwarded
Layer 3 forwarding table	The Layer 3 forwarding table is typically maintained in a TCAM. The forwarding engine examines the Layer 3 forwarding table using the destination IP address of the packet as an index. The longest matching prefix in the table points to the next hop Layer 3 address and outbound port (plus its MAC address) for the packet. The Layer 3 forwarding table also contains the Layer 2 (MAC) address of the receiving interface of the next hop node. Also contained along with the egress port is any VLAN ID for the outgoing packet
QoS ACLs	QoS ACLs are also maintained in TCAMs. In many implementations, information required for packet classification, policing, and marking can all be performed simultaneously in a single lookup in the QoS TCAM
Security ACLs	Inbound and outbound ACLs can also be maintained in a TCAM so that lookups to determine whether to forward or filter a packet can be performed in the TCAM in a single lookup

has to determine the QoS and security handling instructions for the packet by performing lookups in the corresponding ACLs maintained in the TCAM (Figure 11.2). The modules involved in the forwarding of a packet are described in Table 11.2. The forwarding operations are performed in parallel in hardware in some switch and router architectures.

After the Layer 2 or 3 forwarding table lookup, the packet is transferred to the appropriate egress priority queue at the destination switch/router port. The next hop destination information obtained from the Layer 3 forwarding table also comes with a corresponding receiving interface MAC address information. The Layer 3 address (i.e., the IP address in the packet) used to retrieve the next hop and its Layer 2 address may also point to other information regarding tagging, writing, or rewriting certain QoS markings to the departing packet.

The original Layer 2 destination address in the packet (which is the address of the receiving interface of the switch/router) is replaced with the next hop's Layer 2 address. The forwarded packet's Layer 2 source address is changed to that of the switch/router's outbound port before it is transmitted onto the next hop. The Time-To-Live (TTL) value in the Layer 3 packet must be decremented by one.

The Layer 3 packet header checksum must be recalculated because the contents of the Layer 3 packet (the TTL value) have changed. In addition, the Layer 2 checksum must be recalculated because both Layer 2 and 3 addresses have changed. Essentially, the entire Ethernet frame header and trailer must be rewritten before the frame is placed into the egress priority queue. All the above Layer 2 and 3 forwarding operations in addition to the QoS and security ACL operations can be accomplished efficiently in hardware [AWEYA2001, AWEYA2000].

11.4   QoS FEATURES IN THE CATALYST 6500

The QoS features on the Cisco Catalyst 6500 are described below and also illustrated in Figure 11.2. Reference [CISCQOS05] provides a summary of the QoS features and describes where each feature is implemented in the switch/router, whether the feature is on the ingress or egress side of the device. This reference also provides a summary of the QoS capabilities for each of the line cards in the Cisco Catalyst 6500 Family.

Over the years, a number of (different) PFC versions have been developed for the Cisco Catalyst 6500. Reference [CISCQOS05] provides a high-level overview of the major QoS capabilities of each PFC version (PFC1, PFC2, PFC3A, PFC3B, and PFCBXL). PFC4 and related models are not discussed in Ref. [CISCQOS05]. The description of PFC4 and its related models are given in Refs [CISC2TQOS11, CISC2TQOS17].

11.4.1   Packet Classification

Classification is the process or action by which network devices identify specific network traffic or packets so that they can be given a level of service. There are a number of network services and applications where packet classification plays a major role, such as differentiated qualities of service, policy-based routing, network firewall functions, traffic billing, and so on.

Packet classification allows a network device to determine which flow an arriving packet belongs to so that it can determine whether to forward or filter the packet, which port or interface on the device to forward the packet, the class of service (CoS) the packet should be given, how much should the packet and its flow be billed, and so on.

The classification function in network devices is performed by an entity normally referred to as a packet or flow classifier. The classifier stores a number of rules that describe how the classification of packets should take place and also requires that each flow of packets being processed satisfy, at a minimum, one of the stored rules.

The rules maintained by the classifier determine which flow a packet belongs to based on the classifier examining contents of the packet header plus other packet fields. Each rule essentially specifies a class (category, group, or rank) that a packet belongs to based on some criterion derived from the contents of the packet. Also associated with each classification rule is a specified action which is to be carried out when the rule is satisfied.

Classification in the switch/router may be performed based on a number of predefined or selected fields in the arriving packet. For example, a flow could be defined by looking at particular combinations of a packet's source IP address, destination IP address, transport protocol, and transport protocol port numbers. Furthermore, a flow could be simply defined by a specific destination IP address prefix and a range of transport protocol port values, or simply by looking at the port of arrival.

Some examples of classification tools provided by the Cisco Catalyst 6500 are ACLs and per port trust setting on a device [CISCQOS05]. The process of classification in the Catalyst 6500 involves inspecting different fields in the Ethernet (Layer 2) header, along with fields in the IP header (Layer 3), and the Transmission Control Protocol/User Datagram Protocol (TCP/UDP) header (Layer 4) to determine the level of service that will be given to the packet as it transits the switch/router.

11.4.2 Queuing

Queuing in a network device provides a temporary relieve mechanism during short-term congestion by allowing the device to temporarily hold data in memory when the arrival rate of data at a processing point in the device is larger than the departure rate. Network devices use queues to temporarily hold packets until the packets can be processed (forwarding table lookup, classification, packet rewrite, etc.) and forwarded.

Each queue is allocated some buffer memory, which provides the holding space for the data waiting to be processed. In the Catalyst 6500, the number of queues and the amount of buffering allocated to each queue are dependent on the hardware platform used as well as the line card type in use [CISCBQT07].

The Catalyst 6500 Series Ethernet line card modules implement some form of receive (at ingress) and transmit (egress) buffering per port. These buffers are used to store arriving packets as (Layer 2 and 3) forwarding decisions are made within the switch/router, or as packets are being temporarily held for transmission on a port. The buffering becomes more important when the aggregate rate at an output port is greater than the output rate the port supports.

In the Catalyst 6500 architecture, due to the higher switch fabric speed relative to the input ports, the switch fabric itself almost never becomes the bottleneck to data flow to the output ports. Congestion can occur on the transmit (egress) side when more than one port send data to a particular destination port. A majority of the packets entering the switch/router will not experience congestion under light to moderate traffic load conditions. For these reasons, the receive port buffers on the switch/router are configured to be relatively small compared to the transmit port buffers.

When the QoS features are not enabled on the switch/router, all packets have equal access to the port buffers, regardless of the traffic type or class they belong to. Furthermore, in the event of congestion (that is, when a port buffer overflows), all packet regardless of traffic type are equally subject to discard at that port buffer. Packets in the (single queue) buffer are transmitted in the order in which they arrive, and if the buffer is full, all subsequent arriving packets are discarded. This queuing discipline is known as First In, First Out (FIFO) queuing with tail-drop. FIFO queuing is normally used for traffic that requires best-effort service (i.e., service with no guarantees whatsoever).

When the QoS features are enabled on the switch/router, the port buffers are partitioned into a number of priority queues. Each queue is configured with one or more packet discard thresholds. The combination of packet classification, multiple priority queues (within a buffer), and packet discard thresholds associated with each queue allow the switch/router to make intelligent traffic management decisions when faced with congestion. Traffic sensitive to delay and delay variation, such as streaming voice and video packets, can be placed in a higher priority queue for transmission, while other less time-sensitive or less important traffic can be buffered in lower priority queues and are subject to discard when congestion occurs.

11.4.3 Congestion Avoidance

The primary goal of congestion avoidance is managing queue occupancy to avoid overflows and data loss. As a queue accepts data and starts to fill-up during short-term congestion, a congestion avoidance mechanism can be used to ensure that the queue does not fill-up completely. When the queue fills up, subsequent packets arriving to it will simply be discarded, irrespective of the priority class or settings/markings in the packets. Excessive data drops and delays caused by the congestion could affect the performance of some end-user applications.

To address these concerns, congestion avoidance mechanisms are normally used to minimize the potential queue overflows and data loss. Typically, a network device will employ queue occupancy limits or thresholds (lower than the maximum queue size) to signal when certain occupancy levels are crossed. When a queue threshold is crossed, the device can randomly discard lower priority packets while trying as much as possible to accept higher priority packets into the queue. The network device can use congestion avoidance mechanisms such as Random Early Detection (RED) and Weighted Random Early Detection (WRED), which allow for more intelligent packet discard and also perform better than the classical tail-drop mechanism.

When the QoS is enabled on the Catalyst 6500 Series Ethernet line card modules, the multiple priority queues and drop thresholds on the switch/router ports are also enabled. Different priority queues and thresholds can be configured depending on the model of the line card. Each queue can also be configured with one or more packet discard thresholds. The two packet drop threshold types that can be configured are as follows.

11.4.3.1   Tail-Drop Thresholds   On switch/router ports that are configured with tail-drop thresholds, packets of a given CoS value are accepted into the queue until the drop threshold associated with that CoS value is crossed. When this happens, subsequent packets of that CoS value are discarded until the queue occupancy drops below the threshold.

Let us assume, for example, that traffic with CoS = 1 is assigned to Queue 1, which has a queue threshold equal to 60% of the maximum queue size of Queue 1. Then, packets with CoS = 1 will not be dropped until the queue length of Queue 1 is 60% full. When the 60% threshold is exceeded, all subsequent packets with CoS = 1 are dropped until the queue drops below the 60% limit.

11.4.3.2   WRED Thresholds   On switch/router ports configured with WRED thresholds, packet of a given CoS value are accepted into the queue randomly to avoid congestion and queue overflow. The probability of a packet with a given CoS value being discarded or accepted into the queue depends on the thresholds and weight assigned to traffic with that CoS value.

Let us assume, for example, that traffic with CoS = 2 is assigned to Queue 1 with two queue thresholds 40% (low) and 80% (high). Then packets with CoS = 2 will not be dropped at all unless the occupancy of Queue 1 exceeds 40% full. When the queue occupancy exceeds 40% and approaches 80%, packets with CoS = 2 can be dropped with an increasing probability rather than being accepted outright into the queue. When the queue occupancy exceeds 80%, all packets with CoS = 2 are dropped until the queue occupancy falls below 80% once again.

The switch/router drops packets at random when the queue size is between the 40% (low) and 80% (high) thresholds. The packets discarded are selected randomly and not on a per flow basis. WRED is more suitable for traffic generated by rate or window adaptive protocols, such as TCP. A TCP source is capable of adjusting its transmission rate to cope with random packet losses in the network by pausing transmission (backing off) and adjusting its transmission window size to avoid further losses on the transmission path.

Reference [CISCBQT07] describes the structure of the priority queues and thresholds that can be configured on a port in the Catalyst 6500. It describes the number of strict priority queues (when configured) and the number of standard queues along with their corresponding tail-drop or WRED thresholds. The different priority queues and threshold settings on the Catalyst 6500 Ethernet line card modules are also described in Ref. [CISCBQT07].

11.4.3.3   Queue   Configuration   Information   Reference   [CISCBQT07] describes for each of the Catalyst 6500 Series Ethernet line card modules the following queue configuration information:

- Total buffer size per port (total buffer size)
- Overall receive buffer size per port (Rx buffer size)
- Overall transmit buffer size per port (Tx buffer size)

- Port receive queue and drop threshold structure
- Port transmit queue and drop threshold structure
- Default size of receive buffers per queue with QoS enabled (Rx queue sizes)
- Default size of transmit buffers per queue with QoS enabled (Tx queue sizes)

11.4.4 Traffic Policing

Policing is the process of monitoring and enforcing the flow rate of packets at a particular point in a network. The policing mechanism (i.e., the policer) monitors and determines if the rate has exceeded a predefined rate. The rate is determined by measuring traffic within a certain time interval, where, typically, this time interval is a configurable number in the policer. An arriving packet to the policer is deemed to be out-of-profile when it arrives when the flow rate exceeds the predefined rate limit. When this happens, the arriving packet is either dropped or the CoS settings in it are marked down (to a lower priority class).

Traffic policing in the switch/router provides an effective way to limit the bandwidth consumption of traffic passing through a given port or group of ports (as in the case of ports that belong to a single VLAN). Basic policing can be implemented at a port or queue by defining the average rate and burst limit of data that the port or queue is allowed to accept. A policer and a policing policy can be configured that uses an ACL to identify and screen out the traffic that should be presented to the policer.

Multiple policers and policing policies can be configured to work in a switch/router simultaneously, allowing a network manager to set different traffic management policies for different traffic classes passing through the device. A network manager can also configure a policing policy to limit the rate of all traffic entering a particular switch/router port, or the overall rate to a given VLAN by limiting the rate of individual flows to a given rate.

11.4.5 Rewrite

As packets enter or exit a particular network with a given set of CoS offering, border or edge network devices (switches, routers, or switch/routers) might be called upon to alter the original CoS settings of the packets. The network device may be configured with CoS rewrite rules that allow it to rewrite the CoS bit settings within the packet.

Each rewrite occasion requires the network device to read the forwarding (priority) class and loss priority setting within the packet, use this information to lookup the corresponding CoS value from a CoS table containing the rewrite rules, and then write this CoS value over the old value in the packet.

For an arriving packet at a port, an ingress classifier reads the CoS bit setting in the packet and maps that to a forwarding class and packet loss priority combination in a table maintained by the classifier. The device can also be configured to have rewrite rules that allow it to alter CoS bit settings in outgoing packets at the outbound interfaces of the device in order to meet the traffic management objectives of the receiving network. This allows the network devices in the receiving network to receive and correctly classify each packet into the appropriate priority queue with correct bandwidth and packet loss profile.

In addition, an edge device may be required to rewrite a given CoS setting (IP Precedence, Differentiated Services code point (DSCP), IEEE 802.1Q/p, or MPLS EXP (Multiprotocol Label Switching Experimental) bit settings) at its ingress or egress interfaces to facilitate classification, priority queuing, and application of different drop profiles at core or backbone network devices. The Cisco Catalyst 6500 supports CoS rewrite mechanisms that can be used to change the priority value of a packet (increase or decrease it) based on CoS policies that may be configured by the network administrator.

11.4.6 Scheduling

As already discussed, a classifier reads markings or fields in arriving packets and interprets them to be able to assign the packets to different priority queues or process them according to predefined rules. Scheduling is a mechanism used to remove the data in queues and transfer them to other receiving entities. The scheduling mechanism also determines the transmission order of the packets being queued based on the available shared output resources and such that some performance objectives are achieved. The scheduling could be done to satisfy the QoS requirements of each packet or a flow of packets.

The priority queues represent storage locations where the classified packets are temporarily held while waiting to be scheduled and forwarded to the next processing entity. The packet scheduling mechanism often defines precisely the decision process and algorithm used to select which packets should be serviced next or dropped (based on the available system resources and priority markings in the packet). The packet forwarding process may also involve buffer management, which refers to any particular mechanism used to manage or regulate the occupancy of the available memory resources used for storing arriving packets.

When an input or output port is heavily loaded or congested, the FIFO queuing approach may no longer be an efficient way to satisfy the QoS requirements of each flow or user. When supporting different traffic classes, the FIFO approach does not make efficient use of the available system resource to satisfy the different requirements of the classes. Under congestion conditions, multiple packets with different QoS requirements (e.g. bandwidth, delay, delay variation, loss) compete for the finite common FIFO transmission resource.

For these reasons, network devices require classification, priority queuing, and appropriate packet scheduling mechanisms to properly handle the QoS requirements of the different traffic processed, and also the order of packet transmission while accounting for the different QoS requirements of individual packets, flows, or users. In a networking device, scheduling can be done at the ingress and/or egress side of the device.

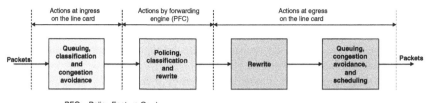

PFC = Policy Feature Card

FIGURE 11.3   Cisco Catalyst 6500 QoS processing model.

11.4.6.1 Input Queue Scheduling   When a packet enters an ingress port, it can be assigned to one of a number of port-based priority queues (based on, for example, CoS marking in the packet) prior to the forwarding table lookup. After the forwarding decision is made, the packet is scheduled and forwarded to the appropriate egress port. Typically, multiple priority queues are used at the ingress port to accommodate the different traffic that require different treatment.

Time sensitive traffic require service levels where latencies in the device (and the network, in general) must be kept to a minimum. For instance, video and voice traffic require low latency, requiring the switch to give priority to these traffic over other traffic from protocols such as File Transfer Protocol (FTP), web, email, Telnet, and so on.

11.4.6.2 Output Queue Scheduling   After the forwarding table lookup and possibly the rewrite processes (see Figure 11.3), the switch/router places the processed packet in an appropriate outbound (egress) priority queue prior to transmission to the outside network. The switch/router may perform congestion management on the output queues to ensure that the queues do not overflow. This is typically accomplished using a mechanism such as RED or WRED, which drops packets randomly (i.e., probabilistically) to ensure that the queues do not overflow.

WRED is a derivative of RED and is used in the Catalyst 6500 family and many other network devices. In WRED, packets are dropped randomly while recognizing the CoS values in the packets. The CoS setting in the packets are inspected to determine which packets will be dropped. When the queue occupancy reaches predefined thresholds, lower priority packets are dropped first to make room in the queue for higher priority packets.

The Cisco Catalyst 6500 supports a number of scheduling algorithms such as strict priority, shaped round-robin (SRR), WRR, and deficit weighted round-robin (DWRR) [CISCQOS05]. The switch/router supports ingress and egress port scheduling based on the CoS setting associated with arriving packets. In the default configuration of QoS in the switch/router, packets with higher CoS values are mapped to higher queue numbers. For example, traffic with CoS 5, which is typically associated with VoIP traffic, is mapped to the strict priority queue, if configured.

12

QUALITY OF SERVICE CONFIGURATION TOOLS IN SWITCH/ROUTERS

12.1 INTRODUCTION

The Catalyst 6500 has a rich set of QoS features and configuration tools that allow it to be used as a good reference platform for understanding the types of configuration tools available for switch/routers. The typical switch/router from other vendors in the market will support these features and configuration tools. Furthermore, devices from different vendors do interoperate very well nowadays partly due to similarities in the capabilities and feature sets built into the devices.

The QoS functions on the Cisco Catalyst 6500 running the Cisco IOS® Software can be configured via the modular QoS command-line interface (MQC) software feature. This toolkit comes as part of the Cisco IOS Software running on Cisco routers and switch/routers. The QoS configuration process can be described by the following steps (Figure 12.1):

1. **Step 1:** Construct a class map that defines the relevant ACLs that identify the particular traffic of interest that QoS is to be applied. The class map defines some traffic classification criteria, essentially it specifies a set of flow matching criteria for the arriving traffic.

   - Class maps can be used to classify traffic based on, for example, Layer 3 and Layer 4 protocol information (source and destination IP address, transport protocol type, source and destination transport protocol port),

Switch/Router Architectures: Shared-Bus and Shared-Memory Based Systems, First Edition. James Aweya.
© 2018 The Institute of Electrical and Electronics Engineers, Inc. Published 2018 by John Wiley & Sons, Inc.

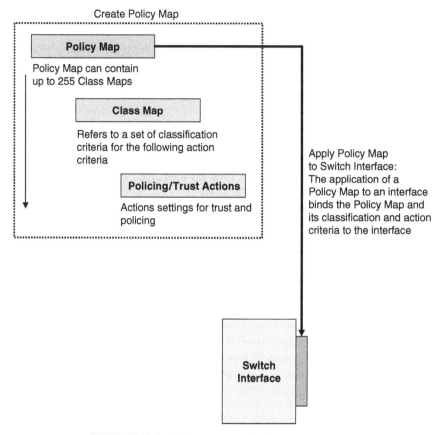

FIGURE 12.1   QoS command-line interface structure.

Layer 7 protocol information (e.g., FTP request commands, HTTP header, HTTP URL, HTTP cookie, HTTP content), and so on.

2. **Step 2:** Create a policy map that references the class map (in Step 1) and defines/specifies a number of actions that are to be performed on the classified traffic. In the QoS context, the policy map includes the QoS policy (priority queuing, bandwidth allocation, drop profile, etc.) that is to be applied to the matched traffic at a switch/router port, group of ports in a VLAN, and so on.

3. **Step 3:** Apply the policy map to the device port, logical interface, a specific VLAN interface, and so on.

Configuring and implementing QoS features on the Cisco Catalyst 6500 [CISC-QOS05] can be greatly simplified through the use of the auto-QoS software tool. Auto-QoS supports a set of QoS software macros that specify a number of QoS functions that can be invoked from the Cisco Catalyst 6500 CLI. Initially, auto-QoS was used to configure QoS features on a given port of a router that supports a Cisco IP Phone.

12.2 INGRESS QoS AND PORT TRUST SETTINGS

A network device (switch, router, switch/router, etc.) can receive a packet at one of its ports or interfaces with a CoS priority value that has already been set by an upstream device. When this happens, it may be important for the receiving device to determine if the CoS priority setting in the arriving packet is valid. In addition to this, the device may want to know whether the CoS value was set by a trusted or valid upstream application or device according to well-established or understood predefined CoS marking rules.

Checking the trustfulness of the CoS settings may be necessary because, for instance, the CoS priority setting may have been carried out by a user hoping to get (unauthorized and unfair) better service from the network. Whatever the reason, the receiving device has to determine if it has to accept the CoS priority value as valid or alter it to another value.

The receiving device uses the port "trust" setting to decide whether to accept the CoS value or not (see Figure 12.2). A port trust setting of "untrusted" results in the receiving device clearing (i.e., wiping out) any CoS priority value carried in an

Trust setting	Result
Untrusted	CoS/ToS set to zero
Trust CoS	CoS/ToS maintained
Trust IP Precedence	CoS/ToS maintained
Trust DSCP	CoS/ToS maintained

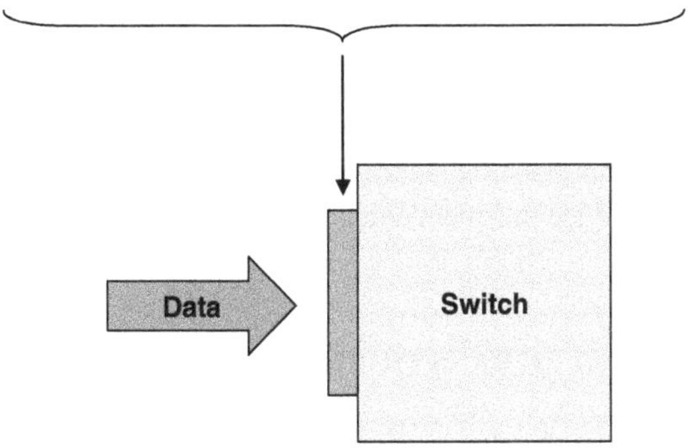

FIGURE 12.2   Switch port trust settings.

arriving packet. The arriving CoS priority setting is not considered trustworthy in this case and has to be rewritten.

The default configuration of QoS on the Catalyst 6500 having all ports on the switch/router is set to the untrusted state. In this configuration, a packet arriving with a CoS priority setting on an untrusted port will have its CoS priority value reset to the default CoS value of zero. The network manager is responsible for identify the ports on the receiving device on which the CoS priority settings of arriving packets should be honored.

For example, the network manager may decide that all connections to well-known and clearly identified servers (e-mail servers, Web servers, etc.), default network gateways, IP telephony call managers, and IP telephones should be set to the trust state where the incoming CoS priority settings are honored. Ports also connected to specific VLANs or IP subnets (such as those carrying management traffic, secured traffic, etc.) can also be preset to the trust state.

A port in the Catalyst 6500 can be configured to trust one of the three priority settings: IEEE 802.1Q/p, IP Precedence, or DSCP. When a port is configured to trust the incoming CoS priority value of packets, the configuration also has to specify which of these three priority setting mechanisms (IEEE 802.1Q/p, IP Precedence, or DSCP) will be trusted.

12.3   INGRESS AND EGRESS PORT QUEUES

The line cards of the Cisco Catalyst 6500 switch/routers support a number of input (receive) and output (transmit) queues per port. These queues are implemented in the port hardware ASICs and are fixed in number (cannot be reconfigured). Each queue is allocated an amount of buffer memory that is used to temporarily store arriving packet on the port. Some line cards have ports that are given a dedicated amount of memory for exclusive use – this memory is not shared with other ports on the line card.

Other line cards have a shared memory architecture where a pool of memory is supported on the card to be shared among a group of ports. The ports are organized in groups, with each allocated a pool of memory. Reference [CISCQOS05] provides a summary list of line card types, their queue structures, and the allocated buffers to each port.

Some line cards have, in addition, a strict priority queue that can be used to queue delay-sensitive and network control traffic. This special queue, typically used for latency-sensitive traffic such as streaming voice and video, is designed to allow the traffic scheduler to service the queued data immediately anytime a packet arrives to this queue.

12.4   INGRESS AND EGRESS QUEUE THRESHOLDS

An important characteristic of data transmission using TCP is that if the network drops a packet, that loss will result in the TCP source retransmitting that packet.

During times of heavy network load and congestion, retransmissions can add to the load the network is already carrying and can potentially lead to buffer overload, overflows, and data loss. To provide a way of managing network load, and ensuring that buffers do not overflow, the Catalyst 6500 switch/routers supports a number of techniques to manage congestion.

Queue thresholds are assigned by the network manager and are predefined queue occupancy limits for a queue. Thresholds define queue fill limits at which the device triggers congestion management mechanisms to start dropping packets from the queue or initial QoS control mechanisms such as packet priority marking, remarking, and so on. Typically, the queue thresholds defined in the device can be used for the functions discussed in the following sections.

12.4.1   Queue Utilization Thresholds

These queue thresholds are used to signal when the buffer space used by a queue has reached a certain predefined occupancy level. When the threshold is crossed, the device will initiate the dropping of newly arriving packets to the queue. The two most common packet drop mechanisms used by network devices are tail-drop and WRED.

12.4.2   Priority Discard Thresholds

These thresholds, typically, are configured on a queue holding packets that can have different priority settings or values. When such a threshold is crossed, a congestion management mechanism such as WRED will start dropping low-priority packets first, and if the congestion persists, the device will move progressively to higher priority packets (Figure 12.3).

These thresholds can also be used to remark packets to lower priority settings when the queue fill exceeds the threshold. The threshold can also be used to trigger when dynamic buffer management mechanisms (for a shared memory buffer pool) can be initiated, for example, to move buffers from ports with lower utilization to ports with high traffic loads.

Furthermore, in load balancing across multiple interfaces (with corresponding queues) on a device, queue thresholds can be used to indicate which interface is lightly loaded and can accept newly arriving flows. A queue (and its interface) that has already crossed its threshold will not be assigned new flows.

12.5   INGRESS AND EGRESS QoS MAPS

The Cisco Catalyst 6500 supports, in addition to class and policy maps, other types of maps that can be used to perform other QoS functions. These maps are described in the following sections.

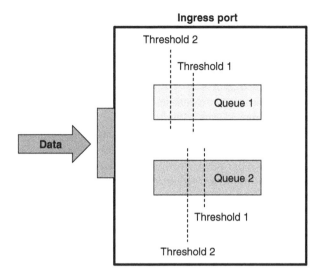

FIGURE 12.3   Mapping a packet to a queue or threshold.

12.5.1   Mapping a Packet Priority Value to a Queue and Threshold

A switch/router port with multiple priority queues requires a mechanism to determine to which priority queue an arriving packet should be placed. The queue placement can be accomplished by using a map that relates the packet's priority setting to a priority queue (Figure 12.3). A classifier decodes/reads the priority setting in an arriving packet and consults the map to determine which of the queues are to store the packet.

This map could be structured or organized to have two columns with the first column storing the possible priority values that can be set in the arriving packets and the second column holding the priority queue (and its associated threshold) to which a packet with a particular priority value should be assigned. It is sometimes necessary to implement an additional default queue to which all other packets, or packets that have markings that cannot fully be interpreted, will be placed.

12.5.2 Mapping Packet Priority Values to Internal Switch Priority Values

A packet that arrives at a switch/router port can already be marked with a priority value by an upstream device. However, the trust setting configured at the port will determine how the priority setting in an arriving packet (IP Precedence, IEEE 802.1p, or DSCP value) will be handled by the switch/router.

When the packet arrives at the switch/router, it is assigned an internal priority value that is of relevance only in the switch/router. The internal priority value is only used in the device for internal QoS management functions and is referred to as the internal DSCP (which is kind of analogous to the DSCP used in DiffServ networks).

The Catalyst 6500 switch/router uses a map to relate the arriving packet's priority setting to an internal DSCP value. A map is used to select an internal DSCP that is predefined for a particular priority setting of incoming packets. After the packet has been processed (forwarding table lookup, priority queuing, etc.) and transferred to the egress port of the switch/router, a second map is used to derive the appropriate IP Precedence, IEEE 802.1p, or DSCP value priority that will be rewritten into the packet before it is transmitted out the egress port. Table 12.1 shows a summary of the maps that can be used by the switch/router. Figure 12.4 shows two examples of the maps (on ingress) that can be used to derive the internal DSCP value.

The IP Precedence is derived from the now deprecated (and now obsolete) IP ToS, which is a 1 byte field that once existed in an IPv4 header. Out of the 8 bits in the IP ToS field, the first 3 bits are used to indicate the priority of the IP packet. These first 3 bits are referred to as the IP Precedence bits and can be set from 0 to 7 in a packet, with 0 being the lowest priority and 7 the highest priority. Cisco IOS has supported setting IP precedence in Cisco devices since many years.

TABLE 12.1 Map Summary

Map Name	Related Trust Setting	Used on Input or Output	Map Description
IEEE 802.1p CoS to DSCP Map	Trust IEEE 802.1p CoS	Input	Derives the internal DSCP from the incoming IEEE 802.1p CoS value
IP Precedence to DSCP Map	Trust IP precedence	Input	Derives the internal DSCP from the incoming IP precedence value
DSCP to IEEE 802.1p CoS Map	—	Output	Derives the IEEE 802.1p CoS for the outbound packet from the internal DSCP

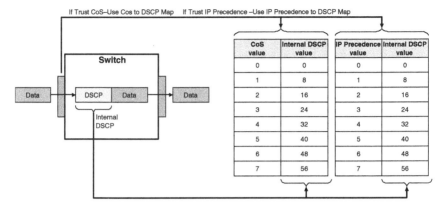

FIGURE 12.4   Mapping priority to internal DSCP.

12.5.3 Policing Map

Policing is primarily employed to limit the traffic rate at a particular point in a network to a predefined rate. Policing can also be used to determine the rate limit at which to lower the priority value of arriving packets when the traffic rate exceeds the rate limit. A switch/router can use a policing map to identify at what rate limit and what priority it will lower the priority settings in arriving packets.

The Catalyst 6500 uses a map called the "policed-dscp-map" to perform the priority remarking task. This policing map is a table that is organized into two columns, with the first column holding the original priority value in a packet and the second column holding the matching value the arriving packet's priority value will be marked down to.

12.5.4 Egress DSCP Mutation Map

As already described, when a packet arrives at a Catalyst 6500 switch/router port, the port's trust setting plays an important role in determining the internal DSCP value to be assigned to the packet. The switch/router uses this internal DSCP value to assign resources to the packet as it passes through the switch/router. When the packet reaches the egress port and before it is transmitted out of the port, a new DSCP value to be written into the outgoing packet is derived from a map using the internal DSCP value as an index (see Figure 12.5).

An egress DSCP mutation map is used to derive the new DSCP value in the outgoing packet. The egress DSCP mutation map contains the information about the DSCP value to be written in the outgoing packet based on the internal DSCP value of the packet. Egress DSCP mutation maps are supported in the Catalyst 6500 with PFC3A, PFC3B, or PFC3BXL modules.

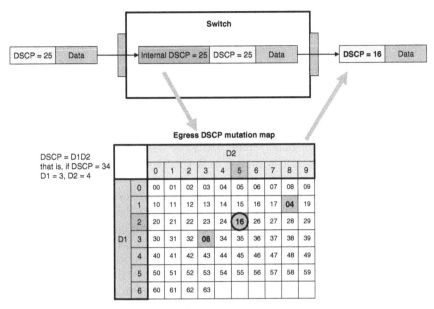

FIGURE 12.5   Egress DSCP mutation.

12.5.5   Ingress IEEE 802.1p CoS Mutation Map

On some Catalyst 6500 line cards, an ingress IEEE 802.1p CoS mutation map can be used on a port that is configured as an IEEE 802.1Q trunk port. This mutation map allows the switch/router to change the incoming IEEE 802.1p CoS value in a packet to another predefined IEEE 802.1p CoS value. An IEEE 802.1p CoS mutation map lists for the possible incoming IEEE 802.1p CoS values corresponding to the outgoing CoS value that can be written in the outgoing packets. A network manager can construct an IEEE 802.1p CoS mutation map to suit the traffic management policy requirements of a particular network.

The IEEE 802.1p CoS mutation map feature is supported on some of the Cisco Catalyst 6500 line cards such as the 48-port GETX and SFP CEF720 Series line card, the 4-port 10GE CEF720 Series line card, and the 24-port SFP CEF720 Series line card. These line cards require a Supervisor Engine 720 to be installed on the Catalyst 6500 chassis for them to function.

12.6   INGRESS AND EGRESS TRAFFIC POLICING

The PFC on the Supervisor Engine and some line card types in the Catalyst 6500 are capable of supporting different policing mechanisms. Policing can be performed on aggregate flows or on microflows passing through the Catalyst 6500 performs. These different policing mechanisms are described in the following sections.

12.6.1 Aggregate Policing

An aggregate policer is a policing mechanism that is used to limit the rate of all traffic that matches a set of classification criteria (defined using an ACL) on a given port or VLAN to a predefined rate limit. The aggregate policer can be applied at a port to rate limit either inbound or outbound traffic. The aggregate policer can also be applied to rate limit traffic in a single VLAN that attaches to multiple ports on a switch/router.

When an aggregate policer is applied to a single port, it meters and rate limits all traffic that matches the classifying ACL passing through the port and the policer. When the aggregate policer is applied to a VLAN, it meters and rate limits all of the matching traffic passing through any of the ports in that VLAN to the predefined rate limit.

Let us assume, for example, that an aggregate policer is applied to a VLAN containing 10 ports on a switch/router to rate limit all traffic to a predefined rate of 50 Mb/s. Then, all traffic entering these 10 ports in the VLAN matching the classification criteria (set by the ACL) would be policed to not exceed 50 Mb/s. The PFC3 on Supervisor Engine 720 can support up to 1023 active aggregate policers in the Catalyst 6500 at any given time.

12.6.2 Microflow Policing

The microflow policer operates slightly differently from the aggregate policer in that it rate limits traffic belonging only to a discrete flow to a predefined rate limit. A flow can be defined as a unidirectional flow of packets that are uniquely identified by IP packet fields such source and destination IP addresses, transport protocol type (TCP or UDP), and source and destinations transport protocol port numbers. The default configuration of microflow policing in the Catalyst 6500 is based on a unique flow being identified by its source and destination IP address, and its source and destination TCP or UDP port numbers.

When applied to a VLAN, the aggregate policer would rate limit the total amount of traffic entering that VLAN at the specified rate limit. The microflow policer, on the other hand, would only rate limit each flow (in a VLAN) to the specified rate. For example, if a microflow policer is applied to a VLAN to enforce a rate limit of 2 Mb/s, then every flow entering any port in the VLAN would be policed to not exceed the specified rate limit of 2 Mb/s. It is important note that although a microflow policer can be used to rate limit traffic for specific flows in a VLAN, it does not limit the number of flows that can be supported or can be active in the VLAN.

Let us consider, for example, two applications – an e-mail client and an FTP session – creating two unique traffic flows through a switch/router port. If a microflow policer is applied to rate limit each one of these flows to 2 Mb/s, then the e-mail flow would be policed to 2 Mb/s and the FTP flow would also be policed to 2 Mb/s. The result of the microflow policing actions produces a total of not more than 4 Mb/s of traffic at the switch/router port.

However, an aggregate policer applied to the same scenario to rate limit traffic to 2 Mb/s would rate limit the combined traffic rate from the FTP and e-mail flows to 2 Mb/s. The PFC on Supervisor Engine 720 supports a total of 1023 aggregate policers and 63 microflow policers.

Another important difference between the microflow and aggregate policers is the location in the Catalyst 6500 where the policer can be applied. On Supervisor Engines 720 and 32 with a PFC3x present, the microflow policer can only be applied at ingress port, while the aggregate policer can be applied at ingress or egress port of the switch/router. Furthermore, in the Catalyst 6500, a microflow policer can support only a single instance of the token bucket algorithm, whereas an aggregate policer can support either a single token bucket or a dual token bucket algorithm.

12.6.3 User-Based Rate Limiting

User-based rate limiting (UBRL) [CISCUBRL06] was first introduced in Supervisor Engine 720 with PFC3 as an enhancement to microflow policing to allow for "macroflow" level policing in a switch/router. UBRL provides a network manager a configuration mechanism to view and monitor flows bigger than a microflow (which are typically defined based on source and destination IP addresses, transport protocol type (TCP or UDP), and source and destinations transport protocol port numbers).

UBRL allows a "macroflow" policer to be applied at a switch/router port to specifically rate limit all traffic to or from a specific IP address (a "macroflow"). In the microflow policing example mentioned earlier, the FTP and e-mail applications create two discrete flows (i.e., microflows) passing through the switch/router port. In this example, each flow is rate limited to the specified 2 Mb/s rate. UBRL slightly enhances the capability of microflow policing in the PFC3 to allow for policing a flow that comprises all traffic originating from a unique source IP address, or destined to a unique destination IP address.

UBRL is implemented as a policer using an ACL entry that has a source IP address only flow mask or destination IP address only flow mask. With this enhancement, a microflow policer can be applied to a switch/router port to limit the traffic rate originating from or going to a particular IP address or virtual IP address. UBRL allows a network manager to configure policing and ACL filtering rules that allow policing based on a per user basis.

The UBRL enhancement is beyond what simple microflow policing can do in the Catalyst 6500. With UBRL and a specified rate limit of 2 Mb/s, if each user (i.e., IP address) in a network initiates multiple sessions (e.g., e-mail, FTP, Telnet, HTTP), all data from that user would be policed to 2 Mb/s. Microflow policing, on the other hand, will rate limit each session to 2 Mb/s, resulting in a total of $N \times 2$ Mb/s maximum rate from the single user, if N sessions are created.

12.7 WEIGHTED TAIL-DROP: CONGESTION AVOIDANCE WITH TAIL-DROP AND MULTIPLE THRESHOLDS

As a queue in a network device begins to fill with data when the traffic load increases or during network congestion, congestion avoidance mechanisms can be used to control data loss and queue overflows. Queue thresholds can be used to signal when to drop packets and what traffic to drop when the thresholds are crossed. Packets can be marked with priority values, and the priority values may indicate to a network device which packets to drop when queue thresholds are breached.

In a multiple threshold single queue system, the priority value in a packet may also identify which particular threshold this packet is allowed to be dropped when the threshold is crossed. When that particular threshold is crossed, the queue will drop packets arriving with that priority value. In such a system, a particular priority value maps to a particular queue threshold value. The queue will continue to drop packets with that priority value as long as the amount of data in the single queue exceeds that threshold. Figure 12.6 illustrates how multiple thresholds can be used in a given queue to selectively discard packets with a particular priority value.

The Catalyst 6000/6500 supports this enhanced version of tail-drop congestion avoidance ("weighted tail-drop") mechanism. This mechanism drops packets marked with a certain CoS priority value when a certain percentage of the maximum queue size is exceeded [CISCQoSOS07]. Weighted tail-drop, allows a network manager to define a set of packet drop thresholds and assign a packet CoS priority value to each threshold.

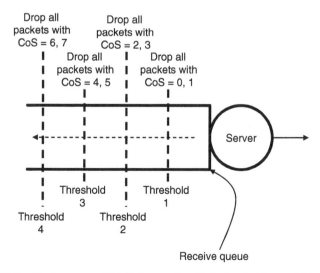

FIGURE 12.6   Single queue with tail-drop and multiple thresholds – weighted tail-drop.

In the following example, we consider a queue that supports four packet drop thresholds. Each drop threshold is defined as follows (Figure 12.6):

- **Threshold 1:** This threshold is set at 50% of the maximum queue size. CoS priority values 0 and 1 are mapped to this drop threshold.
- **Threshold 2:** This threshold is set at 60% of the maximum queue size. CoS priority values 2 and 3 are mapped to this drop threshold.
- **Threshold 3:** This threshold is set at 80% of the maximum queue size. CoS priority values 4 and 5 are mapped to this drop threshold.
- **Threshold 4:** This threshold is set at 100% of the maximum queue size. CoS priority values 6 and 7 are mapped to this drop threshold.

When weighted tail-drop with the above thresholds is implemented at a port, packets with a CoS priority value of 0 or 1 are dropped if the queue is 50% full. The queue will drop packets with a CoS priority value of 0, 1, 2, or 3 if it is 60% full. Packets with a CoS priority value of 6 or 7 are dropped when the queue is completely filled. However, as soon as the queue size drops below a particular drop threshold, the queue stops dropping packets with the associated CoS priority value(s).

When the queue size reaches the maximum configured threshold, all arriving packets are dropped. The main disadvantage of using tail-drop for TCP data transfer is that it can result in a phenomenon generally referred to as "global TCP synchronization." When packets of multiple TCP connections are dropped at the same time, the affected TCP connections go at the same time into the state of TCP congestion avoidance and slow-start to reduce the transmitted traffic. The TCP sources then progressively grow their windows at the same time to cause another traffic peak that leads to further data losses.

This synchronized behavior of TCP transmissions causes oscillations in the data transfer load that shows repeated peak load, low load patterns in the network. Congestion control techniques such as RED and WRED are designed with the goal of minimizing global TCP synchronization, maintaining stable network queues, and maximizing network resource utilization.

12.8   CONGESTION AVOIDANCE WITH WRED

The Catalyst 6500 supports WRED in hardware by implementing WRED in the port ASICs in the line cards. WRED provides a less aggressive packet discard function than tail-drop. WRED spreads packet drops in a queue randomly across all flows in the queue resulting in few flows being severely penalized than the others.

The WRED mechanism works by dropping fewer packets (thereby affecting fewer flows) when it initially starts its drop process. The random and probabilistic drop operations in WRED lead to fewer TCP connections going into "global synchronization" state.

WRED employs multiple thresholds (typically two) when applied to a queue. The lower and upper thresholds are configured to ensure reasonable queue and link utilization and to avoid queue overflow. When the lower threshold is crossed, WRED starts to drop, randomly, packets marked with a particular priority value. WRED tries to spread the packet drops so as to minimize penalizing heavily a few select flows.

As the queue size continues to grow beyond the lower threshold and approaches the upper threshold, WRED starts to more aggressively discard arriving packets. Increasingly, more flows are subject to packet drops because WRED increases the probability of packet drops. The goal here is to signal rate adaptive sources such as those using TCP to reduce their rates to avoid further packet losses and better utilize network resources.

A network device can implement RED or WRED at any of its queues where congestion can occur to minimize packet losses, queue overflow, and the global TCP synchronization problem. The network manager sets a lower threshold and an upper threshold for each queue using RED or WRED and processing of packets in a queue is carried out as summarized below:

- When the queue occupancy is smaller than the lower threshold, no packets are dropped.
- When the queue occupancy crosses the upper threshold, all arriving packets to the queue are dropped.
- When the queue occupancy is between the lower threshold and the upper threshold, arriving packets are dropped randomly according to a computed drop probability. The larger the queue size, the higher the packet drop probability used by RED or WRED.
- Typically, the packet drops are done up to a maximum configured packet drop probability.

If the instantaneous queue size is compared with the configured (lower or upper) queue thresholds to determine when to drop a packet, bursty traffic, potentially, can be unfairly penalized. To address this problem, RED and WRED (and other RED variants) use the average queue size to compare with the queue thresholds to determine when to drop packets. The average queue size is also used in the computation of the drop probabilities used by RED or WRED.

The average queue size captures and reflects the long-term dynamics of the queue size changes and is not sensitive to instantaneous queue size changes and bursty traffic arrivals. This allows the queue to accept instantaneous bursty traffic and also not unfairly penalize them with higher packet losses. This also allows both bursty and nonbursty flows to compete fairly for the system resource without suffering unfair data losses.

RED operates without considering the priority settings in packets arriving at the queue. WRED, however, operates while taking into consideration the priority markings in packets. WRED implements differentiated packet drop policies for packets arriving at a queue with different priority markings based on IP Precedence, DSCP, or MPLS EXP values. WRED randomly drops packets marked with a certain CoS priority setting when the queue reaches a threshold.

With WRED, packets marked with a lower priority value are more likely to be dropped when the lower threshold is crossed. RED does not recognize the IP Precedence, DSCP, or MPLS EXP values in arriving packets. In WRED, if the same packet drop policy (priority unaware policy) is configured at a queue for all possible priority values, then WRED behaves the same as RED.

Both RED and WRED help to mitigate against the global TCP synchronization problem by randomly dropping packets. This is because RED and WRED take advantage of the congestion avoidance mechanism that TCP uses to manage data transmission. RED and WRED avoid the typical queue congestion and data loss that occur on a network device when multiple TCP sessions (as a result of global TCP synchronization) go through the same device port.

However, when RED or WRED is used on the same device port, and when some TCP sessions reduce their transmission rates after packets are dropped at a queue, the other TCP sessions passing through the port can still remain at high data sending rates. Furthermore, because there are TCP sessions in the queue with high sending rates, system resources including link bandwidth are more efficiently utilized. Both RED and WRED provide a smoothing effect on the offered traffic load at the queue (from a data flow and data loss perspective), thereby leading to stable queue size and less queue overflows.

12.9 SCHEDULING WITH WRR

The WRR algorithm can be used to schedule traffic out of multiple queues on each switch/router port where the configuration of the algorithm allows for weights to be assigned to each queue. The weights determine the amount or percentage of total bandwidth to be allocated to each queue. The queues are serviced in a "round-robin" fashion where each queue is serviced in turn, one after the other. The WRR algorithm transmits a set amount of data from a queue before moving to the next queue.

The simple round-robin algorithm will rotate through the queues transmitting an equal amount of data from each queue before moving to the next queue. The WRR, instead, transmits data from a queue with a bandwidth that is proportional to the weight that has been configured for the queue. The amount of bandwidth allocated to each queue when it is serviced by the WRR algorithm depends on the weight assigned to the queue.

The higher the weight assigned to a queue, the higher the bandwidth allocated. The queues with higher weights can send more traffic per scheduling cycle than the queues with lower weights. This allows a network manager to define specific priority queues and configure how much these queues will have access to the available bandwidth. In this setup, the WRR algorithm will transmit more data from specified priority queues than the other queues, thus providing a preferential treatment for the specified priority queues.

12.10 SCHEDULING WITH DEFICIT WEIGHTED ROUND-ROBIN (DWRR)

The deficit round-robin (DRR) scheduling algorithm [SHREEVARG96] was developed as a more effective scheduling algorithm than the simple round-robin algorithm and is used on many switches and routers. DRR is derived from the simpler round-robin scheduling algorithm, which does not service queues fairly when presented with variable packet sizes. With DRR, packets are classified into different queues and a fixed scheduling quantum is associated with each queue. The quantum associated with a queue is the number of bytes a queue can transmit in each scheduling cycle.

The main idea behind DRR is to keep track of which queues were not served in a scheduling cycle (i.e., compute a deficit for each queue) and to compensate for this deficit in the next scheduling round. A deficit counter is used to maintain the credit available to each queue as the scheduling of queues progresses in a round-robin fashion. The deficit counter is updated each time a queue is visited and is used to credit the queue the next time it is revisited and has data to transmit.

The deficit counter for each queue is initialized to a quantum value (which is an initial credit available to each queue). Each time a queue is visited, it is allowed to send a given amount of bytes (quantum) in that round of the round-robin. If the packet size at the head of a queue to be serviced is larger than the size of the quantum, then that queue will not be serviced.

The value of the quantum associated with the queue that was not serviced is added to the deficit counter and will be available as a credit in the next scheduling cycle. To avoid spending valuable processing time examining empty queues (bandwidth wasted), the DRR maintains an auxiliary list called the Active List, which is a list that holds the queues that have at least one packet waiting to be transmitted. Whenever a packet arrives in a queue that was once empty, the index of that queue is added to the Active List.

Packets in a queue visited in a scheduling cycle are served as long as the deficit counter (i.e., the available credit) is greater than zero. Each packet served reduces the deficit counter by an amount equal to the packet's length in bytes. A queue, even if it has packets queued, cannot be served after the deficit counter decreases to zero or negative. In each new round, after a qualified packet is scheduled, from a nonempty queue, the deficit counter is increased by its quantum value.

In general, the DRR quantum size for a queue is selected to be not smaller than the maximum transmission unit (MTU) of the switch or router interface. This ensures that the DRR scheduler always serves at least one packet (up to the MTU of the outgoing interface) from each nonempty queue in the system. The MTU for an Ethernet interface is 1500 bytes but an interface that supports Jumbo Ethernet frames has MTU of 9000 bytes.

12.10.1 Deficit Weighted Round-Robin

Deficit weighted round-robin is a scheduling mechanism used on egress (transmit) queues in the Cisco Catalyst 6500. Each DWRR queue is assigned a relative weight similar to the WRR algorithm. The DWRR scheduling algorithm is an enhanced version of the WRR algorithm. The weights allow the DWRR algorithm to assign bandwidth relative to the weight given to each queue when the interface is congested.

The DWRR algorithm services packets from each queue in a round-robin fashion (if there is data in the queue to be sent) but with bandwidth proportional to the assigned weight, while also accounting for deficits (or credits) accrued during the scheduling cycle. DWRR keeps track of the excess data transmitted when a queue exceeds its byte allocation and reduces the queue's byte allocation in the next scheduling cycles. This way, the actual amount of data transmitted by a queue matches closely the amount defined for it by the assigned weight when compared to the simple WRR.

Each time a queue is serviced, a fixed amount of data are transmitted (proportional to its assigned weight) and DWRR then moves to the next queue. When a queue is serviced, DWRR keeps track of the amount of data (bytes) that were transmitted in excess of the allowed amount. In the next scheduling cycle, when the queue is serviced again, less data will be removed to compensate for the excess data that were served in the previous cycle. As a result, the average amount of data transmitted (bandwidth) per queue will be close to the configured weighted bandwidth.

12.10.2 Modified Deficit Round-Robin (MDRR)

The Catalyst 6500 Series also supports a special form of the DRR scheduling algorithm called modified deficit round-robin, which provides relative bandwidth allocation to a number of regular queues, as well as guarantees a low latency (high priority) queue. MDRR supports a high-priority queue plus regular (unprioritized) queues, while DRR supports only the regular queues. In MDRR, the high-priority queue gets preferential service over the regular queues. The regular queues are served one after the other, in a round-robin fashion while recognizing the weight assigned to each queue.

When no packets are queued in the high-priority queue, MDRR services the regular queues in a round-robin fashion, visiting each queue once per scheduling cycle. When packets are queued in the high-priority queue, MDRR uses one of two options to service this high-priority queue when scheduling traffic from all the queues it handles:

- **Strict Priority Scheduling Mode:** In this mode, MDRR serves the high-priority queue whenever it has data queued. A benefit of this mode is that any high-priority queued traffic always gets serviced regardless of the status of the regular queues. The disadvantage, however, is that this scheduling mode can lead to bandwidth starvation in the regular queues if there is always data queued in the high-priority queue. This can also cause the high-priority queue to consume an unfair and disproportionate amount of the available bandwidth than the regular queues, because this queue is served more often every scheduling cycle.
- **Alternate Scheduling Mode:** In this mode, MDRR serves the high-priority queue in between serving each of the regular queues. This mode does not cause bandwidth starvation in the regular queues because each one of these queues gets served in a scheduling cycle. The disadvantage here is that the alternating serving operations between the high-priority queue and the regular queues can cause delay variations (jitter) and additional delay for the traffic in the high-priority queue, compared to MDRR in the strict priority mode.

For the regular queues, MDRR transmits packets from a queue until the quantum for that queue has been satisfied. The quantum specifies the amount of data (bytes) allowed for a regular queue and is used in the MDRR just like the quantum in the DRR scheduler. MDRR performs the same process for every regular queue in a round-robin fashion. With this, each regular queue gets some percentage of the available bandwidth in a scheduling cycle.

MDRR treats any extra data (bytes) sent by a regular queue during a scheduling cycle as a deficit as in the DRR algorithm. If an extra amount of data were transmitted from a regular queue, then in the next scheduling round through the regular queues, the extra data (bytes) transmitted by MDRR are subtracted from the quantum.

In other words, if more than the quantum is removed from a regular queue in one cycle, then the quantum minus the excess bytes are transmitted from the affected queue in the next cycle. As a result, the average bandwidth allocation over many scheduling cycles through the regular queues matches the predefined bandwidth allocation to the regular queues.

Each MDRR regular queue can be assigned a weight that determines the relative bandwidth each queue receives. The high-priority queue is not given a weight since it is serviced preferentially as described earlier. The weights assigned to the regular queue play an even more important role when the interface on which the queues are supported is congested. The MDRR scheduler services each regular queue in a round-robin fashion if there are data in a queue to be transmitted.

In addition, the Catalyst 6500 Series supports WRED as a drop policy within the MDRR regular queues. This congestion avoidance mechanism provides more effective congestion control in the regular queues and is an alternative to the default tail-drop mechanism. With WRED, congestion can be avoided in the regular MDRR queues by the controlled but random packet drops WRED provides.

12.11 SCHEDULING WITH SHAPED ROUND-ROBIN (SRR)

SRR is another scheduling algorithm supported in the Cisco Catalyst 6500 Series of switch/routers. SRR was first implemented on the uplink ports of the Cisco Catalyst 6500 Series with Supervisor Engine 32 [CISCQOS05]. Unlike WRR, which operates without traffic policing capabilities, SRR supports round-robin scheduling plus a mechanism to shape the outgoing traffic from a queue to a specified rate. The operation of SRR has some similarities to a traffic policer except that data in excess of the specified rate arriving at the SRR scheduler are buffered rather than dropped as in a traffic policer.

The shaper in the SRR scheduler is implemented at the output of each queue and works by smoothing transient traffic bursts passing through the port on which the SRR scheduler is implemented. In SRR, a weight is assigned to each queue, which is used to determine what percentage of output bandwidth the queue should receive. The traffic transmitted from each queue by the SRR scheduler is then shaped to that allocated percentage of the output bandwidth. SRR limits outbound traffic on a queue to the specific amount of bandwidth that its weight allows.

12.12 SCHEDULING WITH STRICT PRIORITY QUEUING

The Cisco Catalyst 6500 also supports strict priority queuing on a per port basis on select line cards. Strict priority queuing is used to service delay-sensitive traffic (like streaming video and voice) that gets queued on the switch/router's line card. In a multiple queue system with a strict priority queue and WRR low-priority queues implemented, when a packet is queued in the strict priority queue, the WRR ceases scheduling of packets from the low-priority queues and transmits the packet(s) in the strict priority queue. The packets from WRR low-priority queues will be served (in a WRR fashion) only when the strict priority queue is empty.

12.13 NETFLOW AND FLOW ENTRIES

NetFlow is a collection of functions used for monitoring and gathering information about traffic that passes through a network device. A switch/router can implement NetFlow as part of an architecture that supports a microflow policer and UBRL to have a better view of the flows these mechanisms are applied to. NetFlow can store information about flows passing through the Catalyst 6500 in memory located on the PFC3 on the Supervisor Engine 720.

12.13.1 NetFlow Entry and Flow Mask

A flow mask is used to define what constitutes a flow and what the NetFlow table stores in a flow entry. The flow mask defines the fields in the arriving packets that identify a flow. It can also be used to define what constitutes a flow in a microflow policer and UBRL.

For example, when used in NetFlow or in the context of UBRL, the following three forms of flow masks can be used [CISCUBRL06]:

- **Source-Only Flow Mask:** The source-only IP address flow mask identifies packets with a particular source IP address in arriving packets as constituting a distinct flow. When a user (i.e., IP address) initiates a Telnet, HTTP, and e-mail session passing through an interface being monitored, traffic from these three separate sessions would be seen as a single flow. This is because the three sessions, although separate, share a common source IP address. Only the source IP address is used as the flow mask to identify unique flows at the monitoring point.

- **Destination-Only Flow Mask:** This flow mask identifies packets with a particular destination IP address as a unique flow. The destination-only IP address flow mask can be used, for example, to identify outbound traffic from an interface to a server (e-mail server, FTP server, Web server, call manager, etc.). This flow mask is used in many cases in conjunction with the source-only IP flow mask.

- **Full Flow Mask:** The full flow mask uses a particular set of source and destination IP addresses, transport protocol type (TCP or UDP), and source and destination port numbers to identify a unique flow. A user who initiates a Telnet and e-mail session would be seen to initiate two separate flows. This is because the Telnet and e-mail sessions will each use distinct destination IP addresses and port numbers that allow them to be identified as distinct flows.

Other applications that use flow entries in a NetFlow table (and flow masks) include Network Address Translation (NAT), Port Address Translation (PAT), TCP intercept, NetFlow Data Export, Web Cache Communication Protocol (WCCP), content-based access control (CBAC), Server Load Balancing, and so on.

A full flow mask is the most specific mask among the three mask types described earlier and its use results in more flow entries being created in a NetFlow table. This can have implications on the processing required for table maintenance in the network device and on the memory requirements for the NetFlow table storage [CISCUBRL06].

The PFC1 and PFC2 (on Supervisor Engines 1 and 2 of the Catalyst 6000/6500) can only run a single flow mask at any one time. When a microflow policer (which operates with a full flow mask) is defined on a PFC1 or PFC2, it requires a full flow mask to be applied. However, as the PFC1 and PFC2 can use only a single flow mask at any one time, this means an active microflow policer will require other processes to use the same full flow mask. This limitation restricts the monitoring capabilities of the PFC1 or PFC2.

Supervisor Engine 720 with PFC3, however, incorporates a number of hardware enhancements over the older PFC1 and PFC2, one of which is the ability to store and

TABLE 12.2   Flow Masks Available on the PFC3x

Flow Mask Type	Description
Source Only	This is a less specific flow mask for identifying flows. The PFC maintains one flow entry for each source IP address identified. All packets from a given source IP address contribute to the information maintained for this entry
Destination Only	This is also a less specific flow mask for identifying flows. The PFC maintains one flow entry for each destination IP address. All packets to a given destination IP address contribute to this entry
Destination-Source	This is a more specific flow mask for identifying flows. The PFC maintains one flow entry for each source and destination IP address pair. All packets between the same source and destination IP addresses contribute to this entry
Destination-Source Interface	This is a more specific flow mask for identifying flows. This flow mask adds the source VLAN SNMP ifIndex to the information in the destination source flow mask
Full	A full flow entry includes the source IP address, destination IP address, protocol, and protocol interfaces. The PFC creates and maintains a separate cache entry for each IP flow
Full-Interface	This is the most specific flow mask for identifying flows. This flow mask adds the source VLAN SNMP ifIndex to the information in the full-flow mask

activate more than one flow mask at any given time. The Supervisor Engine 720 supports a total of four flow masks in hardware [CISCUBRL06]. This capability is available in the PFC3a, PFC3B, and PFC3BXL.

Out of the four flow masks supported, one is reserved for multicast traffic, and another for internal system use. The remaining two flow masks are used for normal operations, such as the masks used in UBRL. The PFC3x also introduces a number of new flow masks as described in Table 12.2 [CISCUBRL06].

12.13.2   NetFlow Table

The information collected for the flow entries in the NetFlow table can be stored in memory in the PFC3x. To facilitate high-speed lookups and updates for flow entries, a TCAM, also located on the PFC, is used to store the NetFlow table. On the Supervisor Engine 720, three PFC3x options (PFC3a, PFC3B, and PFC3BXL) can be supported, each having a different storage capacity. The capacities of each PFC3x with respect to the number of flows that can be stored in the TCAM NetFlow table are described in Ref. [CISCUBRL06].

The PFC uses a hash algorithm to locate and store flow entries in the TCAM NetFlow table. The hash algorithm is used together with the flow mask that

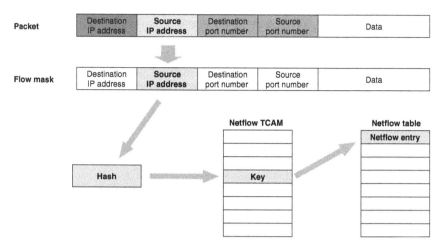

FIGURE 12.7   NetFlow hash operation.

identifies the fields of interest in the arriving packets. The packet fields specified by the flow mask are used as input to the hash algorithm. The hash algorithm output points to a TCAM location, which contains a key. The key in turn provides the index into the NetFlow table, which contains the actual NetFlow entry. This process is illustrated in Figure 12.7.

13

CASE STUDY: QUALITY OF SERVICE PROCESSING IN THE CISCO CATALYST 6000 AND 6500 SERIES SWITCHES

13.1 INTRODUCTION

The key QoS functions in the Catalyst 6000/6500 family of switch/routers requiring real-time processing are implemented in hardware. These QoS-related hardware components are implemented on Catalyst 6000/6500 modules such as the Policy Feature Card (PFC) and the port ASICs on the line cards. This chapter describes the QoS capabilities of the PFC and the switch/router port ASICs on the line cards [CISCUQSC06,CISCUQSC09].

The Multilayer Switch Feature Card (MSFC) supports some QoS functions such as control plane policing and other rate-limiting functions for control and management traffic. These special QoS functions are not discussed in this chapter. Detailed descriptions of the PFC and MSFC are given in Chapters 7 and 9.

13.2 POLICY FEATURE CARD (PFC)

The PFC1 is a daughter card that is supported only on the Supervisor Engine 1A of the Catalyst 6000/6500 family. The PFC2 is an improved design of the PFC1 and is supported on Supervisor Engine 2. PFC1 and PFC2 are the primary modules on which the hardware-based QoS functions are implemented for Supervisor Engine 1 and 2, respectively. Supervisor Engine 32 supports the PFC3B and MSFC2a as a

Switch/Router Architectures: Shared-Bus and Shared-Memory Based Systems, First Edition. James Aweya.
© 2018 The Institute of Electrical and Electronics Engineers, Inc. Published 2018 by John Wiley & Sons, Inc.

default configuration. Supervisor Engines 720, 720-3B, and 720-3BXL all support PFC3 and MSFC3. The newer generations of the PFC support more advanced QoS functions than the older ones.

13.2.1 Policing in the PFC

In addition to some select line card types, the PFC supports virtually all the hardware-based QoS functions, including classification and policing of packets. A PFC can use packet classification together with an access control list (ACL) to mark incoming packets with a priority value. The marking can be based on IEEE 802.1p/Q, IP Precedence, Differentiated Services Code Point (DSCP), or Multiprotocol Label Switching (MPLS) Experimental (EXP) bits. The marked packets then allow other QoS functions to be implemented in the switch/router such as priority queuing and packet discard. We use IEEE 802.1p/Q to designate the priority value in the tagged Ethernet frame in this chapter.

Policing allows a stream of packets arriving at a particular point in the device to be rate limited to a predefined limit. The PFC supports the ability to rate limit (or police) in hardware, incoming traffic to the switch/router to a configurable predefined limit. Packets in excess of the rate limit can be dropped by the PFC or have their priority value marked down to a lower value. A PFC can support ingress and/or egress rate limiting in the switch/router.

In addition to the basic QoS functions of classification, priority queuing, and priority-based packet dropping, the PFC can support additional functions such as the following:

- The PFC can support normal policing where packets are dropped or their priority values marked down if a configured policing policy returns an out-of-profile decision for arriving traffic. The PFC can also support an excess rate policy, where a second policing level can return a policy action on the policed excess traffic.
- The PFC has the ability to push down a QoS policy to a Distributed Forwarding Card (DFC) for it to carry out QoS operations locally. The functions of the DFC are described in Chapter 9.

When an excess rate policer (i.e., a two-rate policer) is defined and applied at an interface, arriving packets can be dropped or their priority values marked down when the excess traffic rate exceeds the predefined excess rate limit. If an excess rate policing level is configured, the PFC uses an internally configured "excess DSCP" mapping table to determine what DSCP value the original DSCP value in a packet should be marked-down to.

If only a normal (single) policing level is configured, a "normal DSCP" mapping table is used. When both excess rate and normal policing levels are configured, the excess rate policing level will have precedence in the PFC for selecting the mapping rules used for priority value mark-downs in arriving packets. A PFC can be configured to support the two types of policing levels at an interface.

The PFC also supports policing at the aggregate flow and microflow levels. These policing functions are described in detail in Chapter 12:

- **Microflow Policing:** In microflow policing, a flow is defined by a unique source and destination IP address, transport protocol type (TCP or UDP), and source and destination port number. For each flow that passes through a port of the switch/router, the microflow policer can be used to limit the rate of traffic received for that flow at the port. Flows that exceed the prescribed rate limit can have their packets dropped or their priority values marked down.

- **Aggregate Policing**: Aggregate policing can be used to rate limit traffic on a port or VLAN that matches a specified QoS ACL. The aggregate flow can be viewed as the cumulative or aggregate traffic at the port or group of ports in a single VLAN that matches a specific Access Control Entry (ACE) of a QoS ACL.

Aggregate and microflow policing provide different ways of specifying the rate of traffic that can be accepted into the switch/router. Both an aggregate and a microflow policy can be configured and made active at the same time at a port or a VLAN. The PFC3 in Supervisor Engine 720 supports up to 63 microflow policers and 1023 aggregate policers.

Policing can be implemented with a token bucket algorithm where the network administrator defines a rate limit and maximum burst size for arriving traffic. The rate limit (or Committed Information Rate (CIR)) is defined as the average rate at which data can be transmitted.

The maximum bust size (or Committed Burst Size (CBS)) is the maximum amount of data (in bits or bytes) that can be transmitted back-to-back to not be out of conformant with the average rate limit. Packets that overflow the token bucket are either dropped or have their priority values marked down.

13.2.2 Access Control Entries and QoS ACLs

A QoS ACL contains of a list of entries (also referred to as ACEs) that define a set of QoS processing rules that can be used by a network device to process incoming packets. An ACE may define traffic classification, priority queuing, discard, marking/remarking, and policing criteria for incoming packets. If the attributes of an incoming packet match the criteria set in the ACE, the receiving device will process the packet according to the "actions" specified by the ACE.

For example, the PFC2 supports up to 500 ACLs, and these ACLs combined can maintain up to 32,000 ACEs in total. The actual number of ACEs the PFC2 can support will depend on the specific QoS functions and services defined on the switch/router and the available memory for ACL storage in the PFC2.

These constraints also apply to other PFC models for the Catalyst 6000/6500 family. The process of creating a policing rule in the PFC or line card involves creating a policer (aggregate or microflow) and then mapping that policer to an ACE in a QoS ACL.

13.3 DISTRIBUTED FORWARDING CARD (DFC)

The DFC allows fabric-enabled and fabric-only (i.e., crossbar switch connected) line cards (see Chapter 9) to perform packet forwarding and QoS processing locally without direct MSFC intervention. In order to maintain consistent operations in the switch/router, the DFC must also support any QoS policies plus ACLs that have been defined in the PFC for the entire switch/router.

In the Catalyst 6000/6500, the QoS policies and related functions cannot be directly configured in the DFC. Instead, the DFC can be programmed with the desired features via the management interfaces in the MSFC and PFC. In a switch/ router configuration with a redundant Supervisor Engine, the DFC can be programmed through the MSFC/PFC on the active Supervisor Engine.

The primary PFC (in a redundant configuration) pushes a copy of its Forwarding Information Base (FIB) to the DFC to allow it to maintain a local copy of the Layer 2 and Layer 3 forwarding tables used for packet forwarding. The PFC will also push a copy of the QoS policies to the DFC so that it can perform QoS operations locally within the line card. With this, the DFC can make local forwarding decisions and also reference the local copy of any QoS policies it has received.

These capabilities allow the DFC to perform QoS processing locally in hardware at wire speeds, thus providing higher overall system forwarding performance. The distributed forwarding and QoS processing functions provided by the DFC allow the switch/router to offload the centralized PFC packet forwarding and QoS operations to the line cards, thereby boosting overall system performance.

13.4 PORT-BASED ASICs

Each line card in the Catalyst 6000/6500 family supports a number of ASICs that implement the queues (along with their associated thresholds) used for temporarily storing packets as they transit the switch/router. The line cards also implement different numbers of queues that are used to prioritize packets at each port.

Each port in turn has a number of input (ingress) and output (egress) queues that are used to temporarily hold packets as they are being processed. The port queues are implemented in the line card hardware ASICs, which hold the necessary memory components. The memory pool is further split up and allocated to the ports on the line card.

Each line card port, typically, has four (congestion management) thresholds configured at each input queue. Each output queue, typically, is assigned two thresholds. These thresholds are also used, during QoS processing, to determine which arriving packet (with a predefined priority value) should be dropped when a particular threshold is crossed.

The network administrator can map different packet priority values to different queue thresholds, signaling to the switch/router which packets to drop when a particular threshold is exceeded. As the queue size builds up and a threshold is crossed, packets arriving with priority values mapped to this threshold will be dropped.

On the Catalyst 6000/6500 family, Weighted Random Early Detection (WRED) packet discarding and weighted round-robin (WRR) scheduling can be based on the priority tag carried in an arriving packet. The tagging enables the switch/router to provide enhanced congestion management with differentiated packet discard and outbound packet scheduling.

A port can employ WRR to schedule outbound traffic from its transmit (Tx) queues to the external network. WRED (and its variants) is a congestion management algorithm employed by the Catalyst 6000/6500 to minimize the impact of dropping packets (while recognizing their priority markings) during times of temporary congestion.

WRED is derived from the RED algorithm and takes into account the priority markings of arriving packets. RED monitors a queue as its occupancy starts to build up and once a predefined threshold has been crossed, packets are dropped randomly based on a computed queue size-dependent probability.

RED is designed to give no preferential treatment to specific flows – All packets arriving to a queue are subject to be dropped randomly. Dropped packets could carry high- or low-priority markings. Also, packets dropped by RED can belong to a single flow or multiple TCP flows. If the packet drops impact multiple flows, then this can have a significant impact on the TCP sources of the flows affected, which in turn reduce their window sizes in response.

Unlike RED, WRED takes into consideration the priority markings of the arriving packets (which could be based on the DSCP, IP Precedence, or IEEE 802.1p/Q priority setting mechanism). In WRED, the network administrator maps drop priority values to specific queue thresholds. Once a particular threshold is exceeded, packets with priority values that are mapped to this threshold are eligible to be dropped.

Packets with priority values mapped to the higher thresholds are accepted into the queue. This prioritized drop and admission process allows higher priority flows to have higher throughput (due to their larger window sizes) and lower latency when the source sends packets to the receiver.

When WRED is not configured on a port, the port uses the tail-drop method of congestion management. Tail-drop simply drops incoming packets once the queue completely fills up.

13.4.1   Original 10/100 Mb/s Ethernet Line Cards (WS – X6348 – RJ45)

The 10/100 Mb/s Ethernet line cards employ a combination of different ASICs to create the modules on which the 48 10/100 Mb/s Ethernet ports reside. Each 10/

100 Mb/s Ethernet line card (root) ASIC provides a breakout facility for 12 10/100 Mb/s Ethernet ports. A line card has four breakout ASICs, each handling 12 10/100 Mb/s Ethernet ports.

The 10/100 Mb/s Ethernet breakout ASIC in turn supports a number of Receive (Rx) and Transmit (Tx) queues for each 10/100 Mb/s Ethernet port. Each breakout ASIC has 128 kB buffering per 10/100 Mb/s Ethernet port. Each port then supports one Rx (ingress side) queue and two Tx (egress side) queues (one designated a high priority and the other as low).

The 128 kB of buffers (per port) is divided between the single Rx queue and the two Tx queues. The single Rx (ingress side) queue is allocated 16 kB of the available memory, and the 112 kB remaining memory is divided between the two Tx queues.

The input queues are used to hold packets, while the port arbitrates for access to the switch fabric. The output queues at a port are used to temporary store packets during high traffic loads to that particular port. The output queues are relatively larger than the input queues because of the potential of many ports sending packets to a single port at any given time.

13.4.2 Newer Fabric 10/100 Mb/s Ethernet Line Cards (WS − X6548 − RJ45)

The newer 10/100 Mb/s Ethernet line card ASICs support a number of Rx and Tx queues for each 10/100 Mb/s Ethernet port. The line card ASICs support a memory pool that is shared by the 10/100 Mb/s Ethernet ports. Each 10/100 Mb/s Ethernet port on the line card in turn supports two Rx queues and three Tx queues. One Rx (ingress side) queue and one Tx (egress side) queue are designated as strict or absolute priority queues.

Packets in these two absolute priority queues are serviced in a strict priority fashion. If a packet arrives in a strict priority queue, the scheduler stops transmitting packets from the lower priority queues to service the packets in the strict priority queue. Only when the strict priority queue is empty will the scheduler recommence servicing packets from the lower priority queue(s). The strict priority queue is normally used as a low-latency queue to handle latency-sensitive traffic such as streaming video and voice traffic.

When a packet arrives at an input or output port during times of congestion, it is placed in one of a number of priority queues. The priority queue in which the packet is placed, typically, is based on the priority value (DSCP, IP Precedence, or IEEE 802.1p/Q) carried in the incoming packet.

At the egress port, a scheduling algorithm is used to service the Tx (output) queues. The scheduling can be done using the WRR algorithm. Each queue is assigned a weight that is used to determine the amount of data to be transmitted from the queue before the scheduler moves to the next queue. The weight assigned to a queue can be configured by the network administrator and is an integer number from 1 to 255.

13.4.3 Gigabit Ethernet Line Cards (WS–X6408A, WS–X6516, WS–X6816)

Each Gigabit Ethernet (GbE) line card has ASICs that support 512 kB of buffering per port. Similar to the 10/100 10 Mb/s Ethernet ports, each Gigabit Ethernet port has one Rx queue and two Tx queues. This queue setup is the default configuration on the WS–X6408–GBIC Gigabit Ethernet line card.

The newer 16-port Gigabit Ethernet line cards, the GBIC ports on the Supervisor Engine 1A and Supervisor Engine 2, and the WS–X6408A–GBIC 8-port Gigabit Ethernet line card support two extra strict priority queues (in addition to the three queues mentioned above). One strict priority queue is assigned as a Rx (ingress side) queue and the other as a Tx (egress side) queue.

These strict priority queues are used primarily for temporarily storing latency-sensitive traffic such as streaming video and voice. With the strict priority queue, packets placed in this queue will be serviced before packets in the high- and low-priority queues. Only when the strict priority queue is empty will the scheduler move to service the high- and low-priority queues.

13.4.4 10 Gigabit Ethernet Line Cards (WS–X6502–10GE)

Each 10 Gigabit Ethernet line card supports only one 10 Gigabit Ethernet port. The 10 Gigabit line card occupies one slot in the Catalyst 6500 chassis. The 10 Gigabit Ethernet line card also supports a number of QoS features. Each 10 Gigabit Ethernet port has two Rx queues and three Tx queues. One Rx queue and one Tx queue are designated as strict priority queues. The Gigabit Ethernet port has a total of 256 kB of Rx buffering and 64 MB of Tx buffering.

13.5 QoS MAPPINGS

The priority value in a packet can be used to determine to which priority queue the packet be placed and at what queue threshold the packet is eligible to be dropped. This is one example of how the Catalyst 6000/6500 uses QoS mappings. When QoS is configured on the switch/router, the following default mappings are enabled:

- The queue thresholds at which packets with specific priority values are eligible to be dropped.
- The priority queue in which a packet is placed based on its priority value.

While the switch/router can use the default mappings, these mappings can be overridden by new settings. Additional mapping can be set such as the following:

- Map IEEE 802.1p/Q priority value in an incoming packet to a DSCP value.
- Map IP Precedence value in an incoming packet to a DSCP value.

- Map DSCP value to an IEEE 802.1p/Q value for an outgoing packet.
- Map IEEE 802.1p/Q priority values to drop thresholds on receive queues.
- Map IEEE 802.1p/Q priority values to drop thresholds on transmit queues.
- Markdown DSCP values for packets that exceed traffic policing limits.
- Set IEEE 802.1p/Q priority value in a packet with a specific destination MAC address.

13.6   QoS FLOW IN THE CATALYST 6000 AND 6500 FAMILY

To facilitate the application of QoS services to traffic passing through the switch/ router, mechanisms to tag or prioritize IP packets or Ethernet frames are required. The IP Precedence and the IEEE 802.1p/Q Class of Service (CoS) bits are two example mechanisms that can be used to perform the tagging.

13.6.1   IP Precedence and IEEE 802.1p/Q CoS Tagging

The IP type of service (ToS) field (now obsolete) consists of 8 bits in the IP packet header. Of these, the first three leftmost bits are used to indicate the priority of an IP packet. These three bits are referred to as the IP Precedence bits. These bits can be set in a packet to take values from 0 to 7, with 0 representing the lowest priority and 7 the highest priority.

The CoS priority in an Ethernet frame can be based on 3 bits in either a Cisco Inter-Switch Link (ISL) header (used to carry VLAN information in Ethernet frames) or an IEEE 802.1Q tag. The latter is the standards-based method used to indicate the priority of an Ethernet frame. The three CoS priority bits in the IEEE 802.1Q tag that can be set in an Ethernet frame are commonly referred to as the IEEE 802.1p bits. The IEEE 802.1Q tag is 4 bytes (32 bits) long.

Of the four bytes, the leftmost 2 bytes are used for the Tag Protocol Identifier (TPID). The remaining 2 bytes are used for the Tag Control Information (TCI). The 16 bit TPID is set to the value of 0×8100 (in hexadecimal) to identify an Ethernet frame as tagged, that is, carrying an IEEE 802.1Q field (tag). The location of the TPID field is at the same position as the type/length field in untagged Ethernet frames. This allows IEEE 802.1Q tagged frames to be quickly distinguished from untagged frames.

The 2 byte TCI field is further divided into a 3 bit Priority Code Point (PCP), 1 bit Drop Eligible Indicator (DEI), and 12 bit VLAN identifier (VID). The 3 bit PCP field is referred to as the IEEE 802.1p CoS and is used to indicate the Ethernet frame's priority level. The DEI bit can be used to indicate an Ethernet frame as eligible to be dropped when a network device decides that frames must be dropped.

Fortunately, the three IEEE 802.1p/Q bits match the number of bits used for IP Precedence. In many networks, there is the need to maintain QoS end-to-end, thereby requiring packets from a user to traverse both Layer 2 and Layer 3 networking domains to a receiver. To maintain QoS end-to-end, the IEEE 802.1p/Q priority values in packets can be mapped to IP Precedence priority values, and IP Precedence priority values mapped to IEEE 802.1p/Q priority values.

Cisco IOS and network devices have supported the setting of IP Precedence for many years. The MSFC or the PFC (independent of the MSFC) in the Catalyst 6000/6500 supports the setting and resetting of the IP Precedence bits in packets. The network administrator can configure a trust setting of "untrusted" at a port that can also wipe out any IP Precedence settings on incoming packets.

More recently, the use of the IP Precedence bits has been included in the capabilities defined by the 6 bits designated as the DiffServ Code Point (DSCP) [RFC2474,RFC2475]. The 6 bit DSCP field results in 64 priority values ($=2^6$) that can be assigned to an IP packet.

The Catalyst 6500 supports the DiffServ's Per Hop Behaviors (PHB) as specified in the IETF standards. The PHB supported include the Default PHB, Class Selector PHBs, Assured Forwarding PHBs, and the Expedited Forwarding PHB. The DSCPs for the PHBs are as follows:

- Assured Forwarding PHB consists of 12 DSCPs.
- Expedited Forwarding PHB DSCP: 101110.
- Default PHB DSCP: 000000.

In addition to defining the Diffserv field in the IPv4 and IPv6 headers, [RFC2474] also defines the Class Selector code points. These code points are the first 3 bits (i.e., the leftmost) of the DSCP that also correspond to the IP Precedence field in the old IP ToS field. These code points (xxx000) correspond to the first 3 bits and with the three rightmost bits all set to 0. The 3 bits result in 8 Class Selector PHBs.

The Catalyst 6000/6500 switch/routers can alter the IP Precedence bit settings in arriving packets. The remarking can be performed using either the PFC or MSFC. When a packet arrives at the switch/router, it is assigned an internal DSCP value. This internal DSCP value is used only within the switch/router to assign different levels of QoS (QoS policies) to arriving packets.

The internal DSCP value may already exist in an arriving packet and be used directly in the switch/router, or the internal DSCP value can be derived from the IEEE 802.1p/Q, IP Precedence, or DSCP value carried in the arriving packet (if the port of arrival is trusted).

The switch/router uses an internal map to derive the internal DSCP. Given that there are eight possible IEEE 802.1p/Q (denoted IEEE here) and eight IP Precedence (denoted IPPrec) values and 64 possible DSCP values, the default map used by the switch/router maps IEEE/IPPrec 0 to DSCP 0, IEEE/IPPrec 1 to DSCP 7, IEEE/IPPrec 2 to DSCP 15, and so on. These default mappings (used by the switch/router) can be overridden by other mappings configured by the network administrator.

When an arriving packet is processed and transferred to an outbound port, the IEEE 802.1p/Q priority value in the outbound packet (to the external network) can be rewritten using a mapping table that translates the internal DSCP value to the new IEEE 802.1p/Q priority value in the packet. The packet exits the switch/router with this new DSCP value on its way to the next node or the final destination.

13.7   CONFIGURING PORT ASIC-BASED QoS ON THE CATALYST 6000 AND 6500 FAMILY

The QoS features configured on the port ASIC by the network administrator can be general enough to affect both inbound and outbound traffic flows in the switch/ router. The following QoS features can be configured:

- Define DSCP to IEEE 802.1p/Q mapping.
- Configure bandwidth on Tx queues.
- Configure priority values to queue threshold mapping.
- Configure Tx queue packet drop threshold.
- Configure Rx queue packet drop threshold.
- Define input (ingress side) classification and port queues priority settings.
- Define the trust state/level of a port.

An Ethernet frame processed by either the MSFC or the PFC is forwarded to the egress port ASIC for further processing. Frames processed by the MSFC have their IEEE 802.1p CoS values reset to zero. Any remarking of priority values has to be done on outbound ports.

Some of the QoS processing performed by the outbound port ASIC (i.e., outbound QoS processing) include the following:

- Assignment of Tx queue tail-drop and WRED thresholds.
- Mapping of IEEE 802.1p CoS values to Tx queue tail-drop and WRED thresholds.
- Remarking of the IEEE 802.1p CoS value in the outbound Ethernet frame using a DSCP to IEEE 802.1p CoS map.

In addition to defining and setting IEEE 802.1p CoS values based on a global port definition, the network administrator can set specific IEEE 802.1p CoS values based on the destination MAC address and VLAN ID. This allows for Ethernet frames destined to specific destinations (e.g., data servers, call manager) to be tagged with a predefined IEEE 802.1p CoS value.

13.7.1   Trust States of Ports: Trusted and Untrusted Ports

The network manager can configure any given port on the Catalyst 6000/6500 switch/ routers as "untrusted" or "trusted." The trust state of a port defines how it classifies, marks (or remarks), drops, and schedules arriving packets as they transit the switch/ router. The default setting of all ports in the switch/router is the untrusted state.

13.7.1.1 Untrusted Ports (Default Setting for Ports) When a port is configured as untrusted, packets entering the port will have their IEEE 802.1p CoS and IP Precedence values reset by the port ASIC to zero. This resetting means the arriving packet will be given the lowest priority service as it transits the switch/router. The network administrator can also configure the switch/router to reset the priority value of any packet that enters an untrusted port to a predefined priority value.

A port set as untrusted will not perform any priority-based congestion management technique such as WRED on its queues. WRED drops packets arriving at a queue based on their priority values once certain predefined queue thresholds are exceeded. All packets entering an untrusted port will be equally eligible to be dropped once the queue is completely full (tail-drop method for congestion management).

13.7.1.2 Trusted Ports A port can be configured to maintain the priority values in packets as they enter the port and transit the switch/router. To allow this, the network administrator can set the trust state of the port as trusted. The untrusted state ignores the priority settings in arriving packets and the switch/router gives them the lowest-priority service.

The switch/outer uses an internal DSCP value to assign a predetermined level of service to the packets arriving through the trusted port. For packets entering a trusted port, the network administrator has to preconfigure the port to look at the existing IEEE 802.1p CoS, IP Precedence, or DSCP priority value in the packet to derive (using a mapping table) the internal DSCP value. The network administrator can also set a predefined internal DSCP value that can be assigned to every packet that enters the trusted port.

For example, the switch/router can use the IEEE 802.1p CoS value carried in the incoming packet to select the internal DSCP. The internal DSCP is then derived using either a default mapping table that was configured when QoS was enabled on the switch/router or a mapping table defined by the network administrator.

13.7.2 Input Classification and Setting Port-Based CoS

On an ingress port of the switch/router, a packet can have its priority value modified if it meets one of the following two criteria:

- The port is configured as untrusted.
- The arriving packet does not have an existing priority value already set.

13.7.3 Configure Receive (Rx) Drop Threshold

At an ingress port, an arriving packet is placed in an Rx queue. To provide congestion control and avoid queue overflows, the port ASIC implements four thresholds on each Rx queue. The port ASIC uses these thresholds to identify which packets can be dropped once these queue fill thresholds have been crossed. The port ASIC can use the priority values in the packets to identify which packets can be dropped when a threshold is exceeded.

As the queue occupancy starts to build up, the occupancy is monitored by the port ASIC. When a threshold is exceeded, packets with priority values predefined by the network administrator are dropped randomly from the queue. This allows higher priority packets to be preferentially accepted into the queue when congestion occurs.

The default drop thresholds can be modified by the network administrator to meet traffic management objectives. Also, the default priority values that are mapped to each threshold can also be modified. Different line cards of the Catalyst 6000/6500 support different Rx queue maximum sizes and thresholds.

13.7.4 Configure Transmit (Tx) Drop Threshold

An egress port may support two Tx queues: Queue 1 and Queue 2, each having two thresholds that are used as part of a congestion management mechanism. Queue 1 could be designated as a standard low-priority queue, while Queue 2 be designated as a standard high-priority queue. Depending on the line card, the congestion management mechanism can be either a tail-drop or a WRED algorithm.

The tail-drop algorithm could employ the two thresholds at a Tx queue in a weighted tail-drop fashion where high-priority packets are dropped only when the higher threshold is crossed (see Chapter 12). In weighted tail-drop, low-priority packets are dropped when the lower threshold is crossed, and both low-priority and high-priority packets are dropped when the higher threshold is crossed.

13.7.5 Mapping CoS to Thresholds

The network administrator can configure thresholds for port queues and then assign priority values to these thresholds. When a threshold is crossed, packets with specific priority values can be dropped. Typically, the network administrator will assign lower priority packets to the lower thresholds, thereby creating space to accept higher priority traffic into the queue when congestion occurs.

13.7.6 Configure Bandwidth on Tx Queues

When a packet is placed in one of the output port queues, it is transmitted out the port using an output scheduling algorithm. The output scheduler configured at the port could be WRR or any derivative of the deficit round-robin (DRR) algorithm (see Chapter 12). Depending on the line card type, a port may support two, three, or four transmit queues.

On the WS–X6248 and WS–X6348 line cards, two Tx queues are serviced by a WRR scheduler. The WS–X6548 line cards support four Tx queues per port. Of these four Tx queues, three are serviced by a WRR scheduler. The fourth Tx queue is a strict priority queue that is always preferentially serviced (over the other three queues) as long as it holds data. The Gigabit Ethernet line cards support three Tx queues, one of which is a strict priority and the other two serviced in a WRR fashion.

Typically, the network administrator assigns a weight to each Tx queue that determines how much traffic will be transmitted from that queue before the WRR scheduler moves to the next queue. A weight from 1 to 255 can be assigned to each of the queues serviced by the WRR scheduler.

13.7.7   DSCP to CoS Mapping

When a packet has been processed and forwarded to an egress port queue, the port ASIC uses the packet's assigned internal DSCP priority value to perform WRED-based congestion management and also to determine the packet's priority queue and its scheduling bandwidth. The switch/router also uses a default map to map back the packet's internal DSCP value to an external IEEE 802.1p CoS, IP Precedence, or DSCP value.

Alternatively, the network administrator can configure a map that can be used by the switch/router to map the assigned internal DSCP value to a new IEEE 802.1p CoS, IP Precedence, or DSCP value for the outbound packet.

13.8   IP PRECEDENCE AND IEEE 802.1p CoS PROCESSING STEPS

Both Layer 2 and Layer 3 switches offer a number of QoS features that include packet classification, input priority queue scheduling, traffic policing and shaping, packet rewriting, and output priority queue scheduling. In general, the classification and QoS features offered by Layer 2 switches are limited to using Layer 2 header information in arriving packets.

The Catalyst 6000/6500 family, on the other hand, have QoS functions processed by a Layer 2 engine that has access to the Layer 3 and Layer 4 packet parameters as well as the to the Layer 2 header information.

As already discussed, in addition to other Layer 2, 3, and 4 packet information, the Catalyst 6000/6500 can access three basic types of packet fields when making QoS decisions [CISCQoSOS07]:

- The IP Precedence that are the first three leftmost bits of the (now obsolete) ToS field in the IP header.
- The DSCP field that are the first six leftmost bits of the newer DiffServ field in the IP header.
- The IEEE 802.1p CoS bits that are 3 bits that are either part of the Cisco ISL header ((used to carry VLAN information in Ethernet frames)) or in the IEEE 802.1Q tag. There is no IEEE 802.1Q tag in an untagged Ethernet frame.

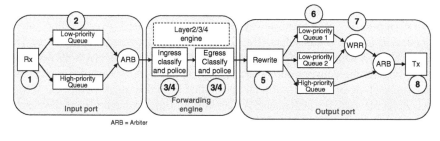

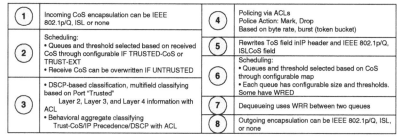

FIGURE 13.1 QoS processing in the Catalyst 6000/6500 switches.

When QoS is disabled in the switch/router, it does not perform any packet classification, marking, or remarking; instead, every packet entering the switch/ router with a DSCP or IP Precedence value leaves with the same priority value (unaltered).

The following sections describe the various QoS operations that are applied to a packet as it transits the switch/router. Figure 13.1 summarizes how the Catalyst 6000/6500 family implements these QoS operations. This figure and the corresponding discussions below describe the QoS operations at the following major modules:

- Input (ingress) port ASIC on a line card.
- Forwarding engine (PFC).
- Output (egress) port ASIC on a line card.

13.8.1 Input Port Processing

The main parameter that affects the QoS configuration (particularly packet classification) at the ingress port is the trust state of the port. Each port of the Catalyst 6000/6500 switch/routers can be configured to have one of the following trust states:

- Untrusted
- Trust-cos

- Trust-ip-precedence
- Trust-dscp

"Trust-cos" refers to setting a port to trust IEEE 802.1p/Q markings in inbound packets. If a port is configured to the untrusted state, an arriving packet is simply marked with the port's default (i.e., predetermined) priority setting and the packet header is sent to the forwarding engine (PFC) for processing.

If the port is configured to have, for example, the trust-cos state, the default port IEEE 802.1p/Q setting is applied if the arriving packet does not already have an IEEE 802.1p/Q or ISL tag. Otherwise, the incoming IEEE 802.1p/Q and ISL tag is kept as it is and the packet is passed to the forwarding engine.

Using IEEE 802.1p/Q as an example, each arriving packet will have an internal DSCP value assigned (either the received IEEE 802.1p CoS or the default port IEEE 802.1p CoS), including untagged Ethernet frames that do not carry any IEEE 802.1Q CoS tags [CISCQoSCM07]. For instance, in a switch/router with a 32 Gb/s shared switching bus (see Chapters 7 and 9), this internal DSCP value and the received IEEE 802.1p/Q value are written in a special packet header (called a Data Bus header) and forwarded over the Data Bus to the PFC (forwarding engine).

The creation and forwarding of the Data Bus header takes place at the ingress line card. Also, at this stage of the packet forwarding, it is not known yet whether the assigned internal DSCP will be carried to the egress port ASIC and written into the outgoing packet. The actual priority value written into the outgoing packet depends on what actions the PFC performs on that packet. The input classification process is described in detail in [CISCQoSCM07].

A packet that enters the switch/router is initially processed by the receiving port ASIC. The port ASIC places the packet in the appropriate Rx queue. Depending on the switch/router line card type, one or two Rx queues may be supported. The port ASIC uses the priority value in the packet to determine which queue to place the packet into (if the port supports multiple input queues). If the port is set as untrusted, the receiving port ASIC can overwrite the existing priority value in the packet with a predefined priority value (Figure 13.2).

13.8.2 Forwarding Engine (PFC) Processing

Once the special packet header has been received by the forwarding engine (PFC), the packet is assigned an internal DSCP. This internal DSCP indicates and associates an internal resource priority to the packet by the PFC as the packet transits the switch/router.

This internal tag is not the same as the normal DSCP value carried in the IPv4 or IPv6 header. It is derived from the existing IEEE 802.1p, IP Precedence, or DSCP setting in the arriving packet and is used as reserve internal switch/router resources and to reset the priority value in the packet as it exits the switch/router. This internal DSCP is assigned to all packets Layer 2 and Layer 3 forwarded by the PFC, and even non-IP packets such as ARP, ICMP, and IGMP messages.

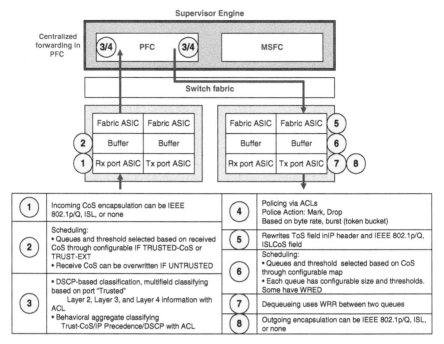

FIGURE 13.2 QoS processing in the Catalyst 6000/6500 switches with centralized forwarding.

The packet is then passed to the Layer 2/Layer 3 forwarding engine in the PFC, which applies any classification and, optionally, traffic policing (rate limiting) policies to the packet. The process of assigning the packet a DSCP value as described above is part of the classification process. This internal DSCP will be used internally by the switch/router for processing the packet until it reaches the egress port. The internal DSCP for a packet will be derived using one of the following:

- The internal DSCP is derived from the IEEE 802.1p CoS value already set in the packet prior to entering the switch/router. Since there are a maximum of eight possible IEEE 802.1p CoS values, each of which must be mapped to one of 64 DSCP values, a mapping table must be used. This mapping table can be created by the network administrator, or the switch/router can use a default map in place.

- The internal DSCP is derived from the IP Precedence value already set in the IPv4 or IPv6 header packet prior to entering the switch/router. As there are only eight IP Precedence values and 64 DSCP values, the network administrator can configure a mapping table that can be used by the switch/router to derive the internal DSCP. A default mapping table can also be used should the network administrator not configure a mapping table.

- The internal DSCP is derived from the existing DSCP value set prior to the packet entering the switch/router.
- The internal DSCP is derived for the packet using a DSCP default value typically assigned though an ACL entry.

The rules that determine which of the above four possible mapping mechanisms to be used for each packet are described in [CISCQoSCM07]. Furthermore, how the internal DSCP is selected depends on the following factors:

- The trust state of the port.
- An ACL applied to the port.
- A default ACL applied to the port.
- Whether an ACL applied is VLAN-based or port-based.

After the PFC assigns an internal DSCP value to the packet, a policing (rate-limiting) policy can then be applied should such a policy be configured. The traffic policing might result in the internal DSCP value of a packet being marked down. The policing may involve the PFC dropping or marking down packets that are out-of-profile. Out-of-profile refers to traffic that has exceeded a rate limit defined by the network administrator (Figure 13.3).

13.8.3 Output Port Processing

After processing the packet, the PFC will then forward the packet to the egress port for final processing. At the egress port, the port ASIC initiates the rewrite process to modify the priority value in the packet. The new value to be rewritten is derived from the internal DSCP. The rewrite can be done according to the following rules:

- If the egress port is configured to perform an ISL or IEEE 802.1Q VLAN tagging, the port ASIC will use an IEEE 802.1p CoS value derived from the internal DSCP, and write this in the ISL or IEEE 802.1Q tagged packet.
- If the packet had an IP Precedence value prior to entering the switch/router, the egress port ASIC will copy the internal DSCP value into the IP Precedence bits of the outgoing header.

This internal DSCP is also used for traffic scheduling at the output port. Once the new priority value is derived from the internal DSCP and written into the packet, the packet is placed in one of the output queues for output scheduling based on its priority value (even if the packet is not IEEE 802.1Q or an ISL tagged).

The packet will then be held temporarily in a transmit queue based on its priority value, ready for transmission. The output port priority queuing can be configured to consist of, for example, one strict priority queue, two standard queues with two thresholds per queue [CISCQoSOS07]. While the packet is in the queue, the port

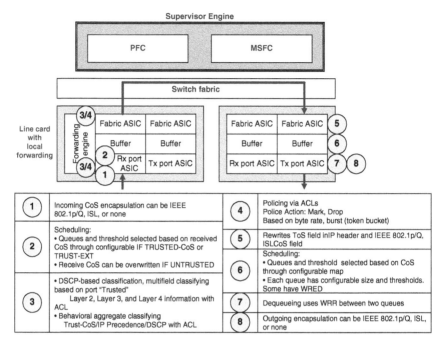

FIGURE 13.3 QoS processing in the Catalyst 6000/6500 switches with distributed forwarding.

ASIC monitors the queue occupancy and implements any WRED actions to avoid buffer overflow.

The output scheduler then selects the queue from which the next packet should be transmitted. A WRR scheduling algorithm could be used to schedule and transmit packets from the egress port queues. The output port also has an arbiter (denoted ARB in Figure 13.1) that checks between each packet transmission from the WRR-serviced queues to determine if there are any data in the strict priority queue that have to be preferentially serviced.

Appendix A

ETHERNET FRAME

A.1 INTRODUCTION

The IEEE standards for Local Area Networks (LANs) describe specifications for the physical layer, medium access control (MAC) sublayer, and logical link control (LLC) sublayer. In OSI terminology, the MAC and LLC sublayers are considered to be sublayers of the OSI Data Link layer. Both the MAC and LLC sublayers contain fields for addressing. The term Ethernet refers to the family of LAN specifications covered by the IEEE 802.3 standard.

The processing that take place in Layer 2 (Ethernet) switches, switch/routers (multilayer switches), and routers involve processing all or parts of the Ethernet frame and IP header fields. To understand how switches, routers, and switch/routers operate, it is essential to understand the format of the Ethernet frame and IP header.

A.2 ETHERNET FRAME FORMAT

The IEEE 802.3 standard defines the basic Ethernet frame format that is required for all Ethernet device implementations, in addition to other optional fields that are used to extend Ethernet's basic capabilities. The basic Ethernet frame contains the Preamble plus the six fields shown in Figure A.1.

Switch/Router Architectures: Shared-Bus and Shared-Memory Based Systems, First Edition. James Aweya.
© 2018 The Institute of Electrical and Electronics Engineers, Inc. Published 2018 by John Wiley & Sons, Inc.

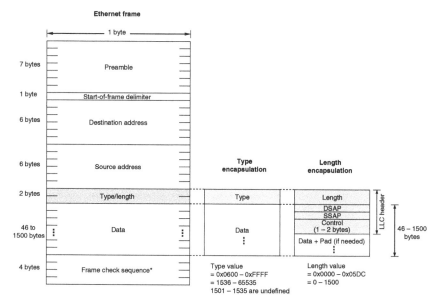

FIGURE A.1   Basic Ethernet frame format.

- **Preamble:** An Ethernet frame is preceded by a Preamble, which consists of 7 bytes[1] or octets (56 bits). The Preamble is a pattern of alternating ones and zeros preceding an Ethernet frame that indicates to a receiving device that an Ethernet frame is arriving. It also provides a signal to the receiving device to enable the frame-reception mechanism of the receiver's physical layer synchronized to the incoming bit stream. The Preamble allows the receiver's physical layer to detect and reach steady-state synchronization with the incoming bit stream (i.e., the Preamble) before the actual Start-of-Frame Delimiter (SFD) of the Ethernet frame is received.

- **Start-of-Frame Delimiter:** The SFD consists of 1 byte field that both marks the end of the Preamble and indicates the beginning of the Ethernet frame. The SFD is a pattern of alternating ones and zeros, ending with two consecutive 1 bits ("11"). It indicates the start of the actual Ethernet frame information and the next bit immediately after the SFD is the left-most bit in the left-most byte of the Ethernet destination MAC address. The SFD has the binary value of 10101011, which is transmitted with the least-significant bit first. Bytes (or octets) in a frame are transmitted least-significant bit first (see bit-ordering and byte-ordering discussion later in the chapter). The purpose of the SFD is to identify the beginning of the MAC frame and allow receiving devices to synchronize with and identify a frame's octet boundaries.

---

[1]The terms "octet" and "byte," both represent groups of eight contiguous bits, and for purposes of this book, the terms are interchangeable.

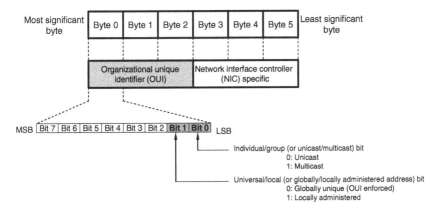

FIGURE A.2   MAC Address format (MAC-48 and EUI-48 formats).

- **Destination Address (DA):** The DA field consists of 6 bytes and it identifies which MAC device(s) the Ethernet frame is destined to. The leftmost bit in the DA field (see Figure A.2) indicates whether the address is an individual address (indicated by a 0) or a group address (indicated by a 1). The second bit from the left indicates whether the DA is globally administered (indicated by a 0) or locally administered (indicated by a 1). The remaining 46 bits of the DA field constitute a uniquely assigned value that identifies a single MAC device, a defined group of MAC devices, or all MAC devices on the network.

  Individual MAC addresses are also known as unicast MAC addresses because they refer to a single MAC address and are assigned by the MAC device manufacturer from a block of addresses allocated by the IEEE. Group addresses (multicast MAC addresses) identify the end devices belonging to a particular group and are assigned by the network manager. Packets sent to a multicast MAC address are received by all MAC devices on a network that has been configured to receive packets sent to that MAC address.

  A special group MAC address (the all 1s or broadcast address) indicates that an Ethernet frame is to be sent to (broadcasted to) all devices on that network. Packets sent to the broadcast address (carrying all ones) are received by all devices on a subnetwork or virtual LAN (VLAN). In hexadecimal format, the broadcast address is FF:FF:FF:FF:FF:FF. A broadcast frame is flooded in a VLAN and is forwarded to and accepted by all other nodes.

- **Source Address (SA):** The SA field consists of 6 bytes and it identifies the MAC devices that send an Ethernet frame. The SA is always an individual MAC address and the left-most bit in the SA field is always 0.

- **Type/Length:** This field consists of 2 bytes and it carries either the Ethernet frame type identifier (ID) (to indicate which protocol is encapsulated in the payload of the frame) or the number of MAC-client data bytes that are contained in the data field of the frame (if the Ethernet frame is created/

assembled using an optional format) (Figure A.1). The Type is sometimes referred to as the EtherType.

- If the Type/Length field value is less than or equal to 1500 (0x05DC), the data field contains LLC data and the number of LLC data bytes carried in the Ethernet frame's Data field is equal to the Type/Length field value.
- If the Type/Length field value is greater than 1536 (0x0600), the Ethernet frame is a Type-frame and the value in the Type/Length field identifies the particular type of frame (upper layer protocol data) being sent or received.
- Values between 1500 and 1536, exclusive, are undefined.

The Type value identifies the upper layer protocol that has encapsulated the data in the Ethernet frame. For example, a Type value of 0x0800 signals that the frame contains an IPv4 datagram. A Type of 0x0806 indicates an ARP frame, 0x8100 indicates an IEEE 802.1Q frame, and 0x86DD indicates an IPv6 frame.

Although both types of framing are formally approved by the IEEE 802.3 standard, the Type frame or encapsulation is the more commonly used one.

- **Data:** The Data field carries a sequence of N bytes of any value, where N is less than or equal to 1500. If the length of data carried in the Data field is less than 46 bytes, the shortfall must be made up, extending the Data field with redundant data (by adding a pad or filler of 0s) to bring the Data field length to 46 bytes. Since the value of the Type/Length field specifies the number of actual/true LLC data octets carried in the Data field (but not the length of the field), the Length value is not changed when a pad is added to the Data field.
- **Frame Check Sequence (FCS):** The FCS field consists of 4 bytes at the end of the Ethernet frame. This sequence contains a 32 bit cyclic redundancy check (CRC) value, which is generated/created by the sending MAC device. The generator polynomial is

$$G(x) = x^{32} + x^{26} + x^{23} + x^{22} + x^{16} + x^{12} + x^{11} + x^{10} + x^8 + x^7 + x^5 + x^4$$
$$+ x^2 + x + 1$$

The FCS is generated over the DA, SA, Type/Length, and Data + Pad fields before transmission. The FCS is recalculated by the receiving MAC device to check for damaged/corrupted frames. The transmitted FCS value in the received frame is compared to the new FCS value that is computed as the frame is being received. The FCS provides error detection over the DA, SA, Length/Type, Data + Pad, and the FCS fields. Note that error detection extends and covers the FCS field itself. The FCS field is transmitted such that the first bit is the coefficient of the x^{32} term (MSB first) and the last bit is the coefficient of the x^0 term. Unlike the byte and bit order in other fields, the FCS is treated as a special 32 bit field rather than as four individual octets.

The Inter-Packet gap (IPG) (not shown in Figure A.1) is the idle (inactive) time between packets. After a packet has been transmitted a sending MAC device, the sender is required to transmit a minimum of 96 bits (12 octets) of idle line state before transmitting the next packet. Between transmission of each Ethernet frame, a sender must wait for a period of 9.6 µs for 100 Mb/s Ethernet and 0.096 µs for Gigabit Ethernet. At 10 Mb/s, this corresponds to about the time it takes for 12 bytes to be transmitted.

The IPG is intended to allow the transmitted signal enough time to propagate through the receiver's electronics at the destination MAC device. While every sending MAC device must pause for the IPG time between transmitting frames, receivers do not necessarily see the IPG or idle/silent period. The time is practically too small to be really noticeable by end applications.

The following describes the frame transmission process for Length encapsulated Ethernet frames. Whenever a sending MAC device receives a transmit-frame request with the relevant destination MAC address and data from the LLC sublayer, the MAC begins the transmission sequence by transferring the LLC data into the MAC frame buffer.

- The preamble data and start-of-frame delimiter data are inserted in the Preamble and SFD fields.
- The destination and source MAC addresses are inserted into the MAC address fields.
- The LLC data bytes are counted, and the number of bytes is inserted into the Length/Type field.
- The LLC data bytes are inserted into the Data field. If the number of LLC data bytes is less than 46, a pad is added to bring the Data field length up to 46.
- An FCS value is generated over the DA, SA, Length/Type, and Data fields (including any pad) and is appended to the end of the Data field as the FCS field.

After the Ethernet frame is completely assembled, the start of actual frame transmission on the transmission medium will depend on whether the MAC is operating in full-duplex or half-duplex mode. Today's Ethernet MAC devices operate in full-duplex mode only.

Ethernet frame reception at the receiving MAC device is the reverse of frame transmission process. The destination MAC address of the received Ethernet frame is parsed and matched against the receiving device's MAC address list (its individual MAC address, its group MAC addresses, and the broadcast MAC address) to determine whether the frame is destined to it.

If a MAC address is matched, the frame length is checked and the receiver's computed FCS is compared to the value in the FCS field (that was generated during frame transmission). If the frame length is correct and the transmitted and computed FCS values match, the receiver determines the frame type from the contents of the Length/Type field. The Ethernet frame's Data field values are then extracted and forwarded to the appropriate upper layer protocol.

A.2.1 MAC Address

Ethernet MAC addresses identify MAC layer entities (physical or logical/virtual interfaces or ports) in networks that implement the IEEE MAC sublayer functions of the data link layer (OSI or TCP/IP reference model). As with most data link addresses, MAC addresses are unique for each MAC port or interface. Ethernet MAC addresses are formatted according to rules set by the IEEE and can be done using one of the following three numbering name spaces (managed by the IEEE): MAC-48, EUI-48 [IEEESAEUI48], and EUI-64 [IEEESAEUI64], where EUI stands for Extended Unique Identifier.

Most Layer 2 networks use one of these three primary numbering spaces – MAC-48, EUI-48, and EUI-64. EUI-48 and EUI-64 identifiers are commonly used to create globally unique interface addresses (also called universally unique MAC addresses). These globally unique addresses are specified in a number of standards that discuss Layer 2 and 3 addressing. Ethernet and ATM interfaces use the MAC-48 address space. IPv6 interface identifiers use the EUI-64 address space.

All three numbering systems use the same format and differ only in the length of the identifier (Figure A.2). The first three bytes in the figure, known as the Organizationally Unique Identifier (OUI), identify the issuer (i.e., organization) of the MAC-48, EUI-48, or EUI-64 identifier [IEEESAOUICID]. These first three byes (which are administered by the IEEE), shown in the figure in transmission order, identify the issuer (assignee or owner), who can be a manufacturer, vendor, or any other organization globally or worldwide.

OUIs are purchased from the IEEE Registration Authority by the issuer/ assignee. The IEEE Registration Authority administers the assignment of these (OUI) identifiers and ensures that they are globally unique to each organization that receives an identifier. The OUIs are used as the first 3 bytes (24 bits) of identifiers in any one the three numbering name spaces (e.g., MAC-48, EUI-48, and EUI-64) to uniquely identify a particular physical or logical interface in a network. When creating an Ethernet MAC address, the OUI is combined with a 3 byte (24 bit) number (assigned by the owner of the OUI) to form the complete Ethernet MAC address. As shown in Figure A.2, the first three bytes of the MAC address are the OUI.

The next 3 bytes in the NIC specific field (for the MAC-48 and EUI-48 formats in Figure A.2) or 5 bytes (for the EUI-64 format) are assigned by the OUI owner (organization) based on their own allocation rules, as long as the addresses are unique. The last 3 (or 5) bytes are typically created using the serial number of the interface, or any other value administered by the OUI owner. MAC-48 and EUI-48 spaces each result in 48 bit addresses, while EUI-64 spaces result in 64 bit addresses, but all three use the same OUI format.

The MAC addresses (with the NIC specific part assigned by the manufacturer of a MAC device) are generally stored in a device's hardware, such as the device's network interface read-only memory (ROM) or some other software or firmware component. The ROM encodes the MAC address (OUI plus NIC-specific number) and is sometimes referred to as the burned-in address (BIA). BIA means the MAC address is stored permanently in the ROM and is copied into an associated random-access memory (RAM) when the interface powers up and initializes. The BIA is also referred to in the network industry by terms such as physical address, hardware address, or Ethernet hardware address (EHA).

In the first byte of the OUI (Figure A.2), the two least-significant-bits of the second set of 4 bits (second nibble) are used as flag bits for some protocol functions. These flag bits can be used to indicate whether the MAC address is an individual (unicast) address or group (multicast) address, or whether a MAC address is universally or locally administered, and so on. As already noted, MAC addresses can either be universally administered or locally administered [IEEESAGMAC].

Universally administered and locally administered addresses are distinguished by appropriately setting the second least significant bit of the most significant byte of the MAC address (indicated by the "Bit 2" bit in Figure A.2). This bit is also referred to as the U/L bit, short for Universal/Local, which identifies how the address is administered.

- **Universally Administered Addresses:** If the "Bit 2" (U/L) bit is 0, the address is universally administered by the IEEE (through the assignment/ designation of a unique owner/company ID (the OUI)). A universally administered address is unique but the NIC-specific part is assigned to a device by its manufacturer.
- **Locally Administered Addresses:** If the "Bit 2" (U/L) bit is 1, the address is locally administered. A locally administered address is assigned to a device by a network administrator, overriding any burned-in address. Locally administered addresses do not contain OUIs.

The least significant bit of the most significant byte of an address ("Bit 0" bit in Figure A.2) indicates whether an address is an individual (unicast) address or group (multicast) address [IEEESAGMAC]. This bit is also referred to as the I/G bit, short for Individual/Group, which identifies how the address is used.

- **Individual (Unicast) Address:** The "Bit 0" (I/G) bit is set to 0 in individual addresses; meaning if the least significant bit of the most significant octet of an address is set to 0 (zero), the frame is destined to only one receiving interface (unicast transmission).
- **Group (Multicast) Address**: The "Bit 0" (I/G) bit is set to 1 in group addresses. If the least significant bit of the most significant address octet is set to 1, the frame is destined to more than one receiver (group of receivers). The sender transmits the frame only once, and a receiver accepts the frame based on a configurable list of multicast MAC addresses it maintains. Group addresses include broadcast and multicast addresses, and not unicast or individual addresses. Group addresses, like individual addresses, can be universally or locally administered.

A.2.1.1 MAC-48 and EUI-48   The EUI-48 is a 48 bit identifier defined by the
IEEE as a concatenation of a 24 bit OUI value administered by the IEEE Registration Authority, and a 24 bit extension identifier assigned by the organization who
owns that OUI (the organization that purchased the OUI). MAC-48 addresses
(obsoleted by the term EUI-48) are the most commonly used MAC addresses in
most networks. These addresses are generally expressed in the format of 12-digit
hexadecimal numbers (48 bits in length) and can be written using one of the
following formats:

• MM:MM:MM:SS:SS:SS
• MM-MM-MM-SS-SS-SS

The first three octets (MM:MM:MM or MM-MM-MM) are the OUI, which is the
ID of the OUI owner, for example, a hardware manufacturer. As already stated, the
OUI values (manufacturer ID numbers) are assigned by the IEEE. The last three
octets (SS:SS:SS or SS-SS-SS) constitute an ID assigned by the OUI owner, for
example, the serial number for a device, which is assigned by the manufacturer. For
example, an Ethernet interface card might have a MAC address that is of form
00:14:22:01:23:45 (where 00:14:22 is the OUI and 01:23:45 is OUI owner
assigned).

Another method for creating addresses is to use Individual Address Block (IAB)
identifiers. An IAB-based identifier is created by combining a 24 bit OUI (owned
and managed by the IEEE Registration Authority), a 12 bit extension identifier (also
assigned by the IEEE to identify an assignee/owner), and a 12 bit value provided by
the owner to identify individual devices. This method results in unique 48 bit
identifiers (addresses) that also identify the assignee/owner of the IAB. The method
provides only 4096 unique EUI-48 identifiers (addresses) for use by the IAB owner.

For example, if the IAB base value assigned by the IEEE is XX-XX-XX-XX-X0-
00 and the 12 bit extension identifier provided by the IAB owner is ccc (in
hexadecimal), then the EUI-48 identifier created by concatenating these two
numbers is XX-XX-XX-XX-Xc-cc. When an organization requires no more than
4097 unique 48 bit EUI-48 identifiers, then an IAB is ideal for that organization.

There is no real distinction between EUI-48 and MAC-48 identifiers, the only
difference lies in the way the two terms are used in the communication industry.
MAC-48 was used to refer to addresses assigned to network hardware interfaces.
EUI-48 (which is the preferred term, and is used more broadly) refers to identifiers/
addresses that identify a hardware device instance, hardware interface/port, virtual/
logical interface, software interface, model number for a product, form/function of
vendor-specific content, or any other object that requires unique identification.

An EUI-48 can be used as an identifier for a wide range of hardware and software
entities and does not necessarily have to be a network address. An EUI-48 identifier
encompasses more, and is not in fact a "MAC address," although both have formats
that are indistinguishable from each other and are assigned from the same
numbering space.

The label MAC-48 is now considered to be obsolete by the IEEE and the term EUI-48 is used instead. MAC-48 was used to refer to specific type of EUI-48 identifiers used for addressing hardware interfaces in IEEE 802-based technologies such as IEEE 802.3 (Ethernet), IEEE 802.4 (Token Bus), IEEE 802.5 (Token Ring), and IEEE 802.6 (FDDI). This 48 bit address space also contains potentially 2^{48} or 281,474,976,710,656 possible addresses.

A.2.1.2 EUI-64 The EUI-64 identifier, defined by the IEEE, represents a newer standard for creating identifiers similar to the EUI-48. An EUI-64 is a 64 bit identifier created by concatenating the 24 bit OUI value (administered by the IEEE Registration Authority), and a 40 bit extension identifier assigned by the OUI owner. The OUI (carrying the organization ID) is still 24 bits (as in the EUI-48), but the OUI owner assigned extension ID is 40 bits, resulting in a much larger address space for the organization. The EUI-64 identifier uses the U/L and I/G bits in the same way as described above and in Figure A.2.

The IEEE guidelines for creating EUI-64 identifiers [IEEESAEUI64] does not allow the first four nibbles (or 2 bytes) of the extension identifier (i.e., first four nibbles of the organizationally assigned identifier portion of an EUI-64) to be $FFFE_{16}$ or $FFFF_{16}$. This means EUI-64 identifiers of the form cccccFFFEeeeeee and cccccFFFFeeeeee should not be used. This restriction is to allow the encapsulation of EUI-48 values into EUI-64 identifiers.

Similar to EUI-48 identifiers, EUI-64 identifiers can be used to identify various hardware and software instances of a device or any other object that requires unique identification, regardless of application (e.g., network interfaces, software interfaces, virtual/logical interfaces, wireless devices, IEEE 1588 clocks, sensors, etc.). EUI-64 identifiers are used in IEEE 1394 (FireWire), wireless personal-area networks (based on 6LoWPAN (RFC 4944), IEEE 802.15.4, or ZigBee), and IPv6 (using the Modified EUI-64 format as the least-significant 64 bits of a unicast network address or link-local address when stateless autoconfiguration is used).

MAC-48 or EUI-48 identifiers can be embedded in EUI-64 identifiers by a simple mapping mechanism. With this mapping, a MAC-48 or EUI-48 identifier can be encapsulated within the larger EUI-64 identifier. The EUI-64 identifier is created by combining the smaller EUI-48 or MAC-48 identifier with specified values written in specified bit locations within the EUI-64 identifier [IEEESAEUI64]. The mapping requires the insertion of the hexadecimal value FF-FF for MAC-48 to EUI-64 mapping, and the value FF-FE for EUI-48 to EUI-64 mapping.

This mapping allows for a smooth transition from MAC-48 and EUI-48 to EUI-64 and to avoid duplicate or conflicting values from occurring during the conversion of MAC-48 and EUI-48 identifiers to EUI-64.

- **MAC-48 to EUI-64:** To convert a MAC-48 identifier into an EUI-64, copy the 24 bit OUI from the MAC-48 identifier, then attach the identifying sequence FF-FF, and then followed by the 24 bit extension identifier specified by the OUI owner.

- **EUI-48 to EUI-64:** To convert an EUI-48 identifier into an EUI-64, the same process is followed, but instead the special identifying sequence FF-FE is inserted.

In both mapping mechanisms, reversing the process, when necessary, is simple and straightforward. Organizations that issue EUI-64 identifiers have to take the necessary measures to guard against issuing identifiers that could be confused with these two mapping forms. The IEEE policy favors the use of the EUI-64 numbering name space when possible over new uses of MAC-48 and EUI-48 identifiers.

IPv6 uses a Modified EUI-64 in the lower half of some IPv6 addresses. A Modified EUI-64 is an EUI-64 identifier with the U/L bit inverted [RFC7042]. To create the Modified EUI-64, IPv6 uses EUI-48 instead (not MAC-48) and also toggles the U/L bit. To create an EUI-64 identifier from an EUI-48 identifier, the 16 bit sequence FF-FE (11111111 11111110) is inserted into the EUI-48 identifier between the 24 bit OUI and the 24 bit extension identifier.

Then to obtain the 64 bit interface identifier for an IPv6 unicast address, the U/L bit in the EUI-64 identifier just created is complemented, that is inverted (if it is set to 1, it is changed to 0, and if it is set to 0, it is changed to 1). This allows MAC addresses (such as EUI-48 formatted IEEE 802 MAC addresses) to be extended to Modified EUI-64 using only the special identifying sequence FF-FE (and never FF-FF) and with the U/L bit inverted.

A.2.1.3 Written Address Conventions The following are standard formats for writing MAC-48 and EUI-48 addresses so that they are easily readable (all in transmission order):

- **Six Groups of Two Hexadecimal Symbols, Separated by Hyphens (–):** Example, 01-23-45-67-89-ab
- **Six Groups of Two Hexadecimal Symbols, Separated by Colons (:):** Example, 01:23:45:67:89:ab
- **Three Groups of Four Hexadecimal Symbols Separated by Dots (.):** Example, 0123.4567.89ab

The first two forms are also commonly used for EUI-64 identifiers.

A.2.2 IEEE 802.1Q Tagging Option – VLAN Tagging Option

The Ethernet frame header contains destination and source MAC addresses (each 6 bytes in length), the Type/Length field, and, optionally, an IEEE 802.1Q tag (Figure A.3). The optional 4-octet IEEE 802.1Q tag is a field that is used to carry information about the Virtual LAN (VLAN) membership and IEEE 802.1p priority of a frame. The minimum payload of the Ethernet frame is 42 bytes when it carries an IEEE 802.1Q tag and 46 bytes when the tag is absent (untagged frame). The maximum Ethernet frame payload, as discussed above, is 1500 bytes. Nonstandard

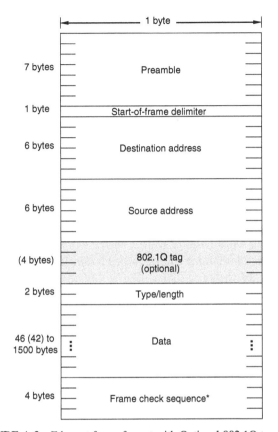

FIGURE A.3 Ethernet frame format with Optional 802.1Q tag field.

jumbo frames are larger size frames, carrying payloads much larger than 1500 bytes, and allow for larger maximum payload size to be transported.

VLAN tagging (using IEEE 802.1Q tags) is a MAC frame formatting option that provides three important capabilities not previously available to Ethernet network users and network managers using untagged frames:

- VLAN tagging provides a means to tag frames with priority settings according to traffic type and quality of service (QoS) requirement so that they can be given differentiated service in a network.

- VLAN tagging allows frames from end stations to be tagged and assigned to Layer 2 broadcast domains or logical groups (VLANs) in a network. Users, even if not collocated within a logical group, can communicate with each other across the network as if they were on a single LAN. Layer 2 switches will examine destination MAC addresses and VLAN tags and forward frames only to ports that serve the VLAN to which the traffic belongs.

- VLAN tagging simplifies network management and makes adding, moving, and changing end-stations and network equipment in the network much easy to carry out.

As shown in Figure A.3, a VLAN-tagged frame is simply a basic MAC frame that has a 4 byte VLAN tag/header (IEEE 802.1Q tag) inserted/placed between the source MAC address and Type/Length fields.

The 32 bit IEEE 802.1Q tag field is placed between the source MAC address and the Type/Length fields of the basic frame (Figures A.4–A.6). The IEEE 802.1Q tag consists of the following parts:

- **Tag Protocol Identifier (TPID):** The 2 byte TPID field is set to a value of 0x8100 (in hexadecimal) to identify an Ethernet frame as carrying an IEEE 802.1Q-tagged. As illustrated in Figure A.4 Figure A.5 Figure A.6, the TPID field is at the same position as the Type/Length field in untagged Ethernet frames. This allows IEEE 802.1Q tagged frames to be quickly distinguished from untagged frames.

- **Tag Control Information (TCI):** The 2 byte TCI field consists of a 3 bit Priority Code Point (PCP) (or User Priority), 1 bit Drop Eligible Indicator (DEI) (formerly the Canonical Format Indicator (CFI)) and 12 bit VLAN Identifier (VLAN ID) as shown in Figure A.5.
  - **Priority Code Point (PCP):** The 3 bit PCP (or User Priority) field refers to the IEEE 802.1p priority class of service and indicates the Ethernet frame's

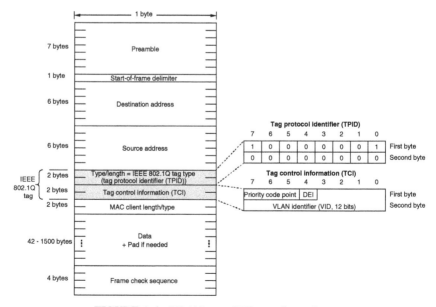

FIGURE A.4   VLAN tagged Ethernet frame format.

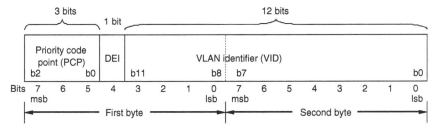

DEI = Discard eligibility indicating

FIGURE A.5 VLAN tag TCI (TAG Control Information Field) format.

6 bytes	6 bytes	2 bytes	2 bytes	2 bytes	42–1500 bytes	4 bytes
Destination address	Source address	TPID =0x8100	TCI	Type/ length	Data	FCS

FIGURE A.6 VLAN Protocol Identifier (VPID) on Ethernet.

priority level. These PCP values can be used to prioritize different classes of traffic (voice, video, data, etc.).

- **Drop Eligible Indicator (DEI):** The 1 bit DEI field (formerly designated as the CFI) can be used separately or in conjunction with PCP to indicate an Ethernet frame as eligible to be dropped when the need arises in the network.

- **VLAN Identifier (VLAN ID):** The 12 bit VLAN ID field carries a value that specifies the VLAN to which the Ethernet frame belongs.

In the range of values covered by the 12 bit VLAN ID field (Figure A.4 and A.5), the values 0x000 and 0xFFF (in hexadecimal) are reserved. The remaining ($2^{12} - 1$) values are available to be used as VLAN identifiers, meaning up to 494 VLANs can be created.

- **The Null VID (0x000):** The reserved value 0x000, when carried in a tagged Ethernet frame, indicates that the frame does not belong to any VLAN (no valid VLAN identifier is carried in the frame). It indicates that the IEEE 802.1Q tag header (i.e., the TCI) contains only frame priority information (in PCP and DEI fields). The 0x000 VLAN ID value is not to be used as a Port VLAN ID, member of a VLAN ID set, configured in any Ethernet address filtering database entry, or used in any management operation.

- **The Default VID (x001):** In Layer 2 networks, the 0x001 VLAN ID or VLAN 1 (normally designated the default VLAN ID) is often reserved for identifying Ethernet frames belonging to a management VLAN, but its use is also vendor-specific.

- **Reserved for Implementation Use (0xFFF):** The 0xFFF VLAN ID value is reserved and is not to be used as a Port VLAN ID, member of a VLAN ID set,

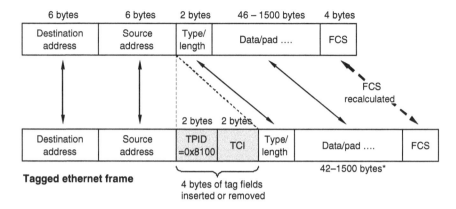

FIGURE A.7   Mapping between untagged and tagged Ethernet frame formats.

or carried in an IEEE 802.1Q tag of an Ethernet frame. This VLAN ID value is sometimes used to indicate a wildcard match in the VLAN ID field in Ethernet address filtering database entries or in searches involving the VLAN ID in management operations.

As illustrated in Figure A.7, IEEE 802.1Q tagged frame has a 4 byte field added between the source MAC address and the Type/Length fields of the basic standard Ethernet frame. With tagging, the minimum frame size remains unchanged at 64 bytes (octets) but the maximum frame size extends from 1518 to 1522 bytes. With IEEE 802.1Q tagging, the minimum payload in the frame is 42 bytes while without tagging the frame carries the basic standard minimum payload of 46 bytes. During tagging, a device adds 2 bytes for the TPID and 2 bytes for the TCI. The device also has to fill in the TCI fields (PCP, DEI, and VID) with the appropriate information. By inserting the IEEE 802.1Q tag, the frame size and contents change, thereby requiring the device to recalculate and update the FCS field in the Ethernet trailer.

The Maximum Transmission Unit (MTU) is the size (in bytes) of the largest protocol data unit (PDU) that a particular protocol layer can forward to the next entity. The Ethernet frame size specifies the size of the complete Ethernet frame, including the header and the trailer, but the MTU size in Ethernet refers only to the Ethernet payload. The standard Ethernet frame (not Jumbo frames) has an MTU of 1500 bytes (Figure A.7). The MTU does not include the Ethernet header plus the Cyclic Redundancy Check (CRC) trailer (which is carried in the Frame Check Sequence (FCS) field) in the untagged frame. The header plus the trailer in the untagged frame are 18 bytes in length, resulting in total Ethernet frame size of 1518 bytes.

"Baby giant" frames is a term that refers to Ethernet frames with MTU sizes greater than the standard maximum MTU size up to 1600 bytes. "Jumbo" frames is used to refer to Ethernet frames with MTU sizes up to 9000 bytes [CISCBABY, ETHERALL09]. By extending the maximum Ethernet frame size from 1518 to 1522 bytes due to IEEE 802.1Q tagging (to accommodate the 4 byte tag), some network devices that cannot process tagged frames simply discard such frames. Some network devices that do not support processing for these larger tagged frame size will process and forward the tagged frame successfully, but may flag them as malformed "baby giant" packets [CISCBABY].

A network can be configured to have segments that are VLAN-aware (i.e., IEEE 802.1Q enabled), where frames carry VLAN tags, and segments that are VLAN-unaware (i.e., IEEE 802.1D enabled), where frames do not include VLAN tags. A VLAN tag is added to a frame when it enters the VLAN-aware segment of the network, (see Figure A.7) to specify/indicate the VLAN membership of the frame. Each frame must be recognizable and distinguishable as belonging to precisely only one VLAN. A frame in the VLAN-aware segment of the network that does not carry a VLAN tag is assumed to belong to the native (or default) VLAN.

The 12 bit VLAN ID field in the IEEE 802.1Q tag allows a theoretical maximum of 4096 VLANs to be supported (4094 in practice, taking away the 0x000 and 0x001 VLAN IDs). While this maximum number may be adequate for most smaller networks, there are many networking scenarios where double-tagging (IEEE 802.1ad, also known as provider bridging, Stacked VLANs, or simply QinQ or Q-in-Q) need to be supported. Double-tagging can be useful for large networks and Internet service providers, allowing them to support a larger number of VLANs, in addition to other important benefits. Double-tagging can support a theoretical maximum of $4096 \times 4096 = 16,777,216$ VLANs.

A.2.3   Ethernet Byte and Bit Ordering

The endianness of a network protocol defines the order in which the protocol sends and receives bits and bytes of an integer field of protocol data. In a big endian system, the most significant byte (or bit) is transmitted first. By contrast, in a little endian system, the least significant byte (or bit) is transmitted first. The little endian and big endian bit ordering are illustrated in Figure A.8.

Lower layer protocols, such as Ethernet, Token Ring, FDDI, and ATM, define the order in which bit/byte transmission/reception should occur, and in some of these lower layer protocols the order is the reverse of that of the supported upper-layer protocol. In Ethernet transmission, the byte order is big-endian, that is, leftmost or most significant byte is sent first.

However, the bit order of Ethernet is little endian, that is, the bit corresponding to the 2^0 numerical position or LSB (least significant bit) of the byte is sent first, and the bit corresponding to the 2^7 numerical position or MSB (most significant bit) is sent last. Figure A.9 shows the byte ordering and bit ordering in the Ethernet frame. The

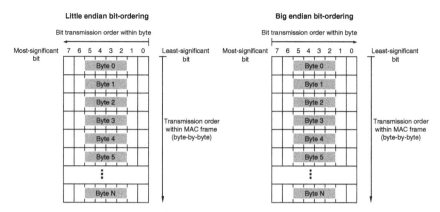

FIGURE A.8   Bit ordering.

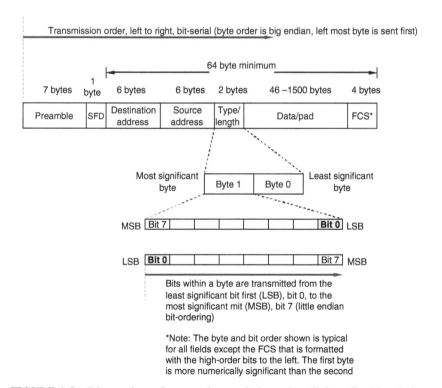

FIGURE A.9   Ethernet frame format and transmission order – little endian bit ordering.

bytes in each field are transmitted in big endian, meaning "left byte first," but the bits within each byte are transmitted in the reverse order "LSB first."

IEEE 802.3 (Ethernet) (and IEEE 802.4 (Token Bus)) transmit the bytes (octets) over the transmission medium, left-to-right, with least significant bit in each byte first (Figure A.9). Also by convention, when writing binary fields, the LSB is shown as the leftmost bit and the MSB as the rightmost. This is also the standard notation (also called canonical format) for MAC addresses, that is, addresses are written in transmission bit order with the LSB transmitted first.

For example, an address in canonical form 12-34-56-78-9A-BC would be transmitted over the transmission medium as bits 01001000 00101100 01101010 00011110 01011001 00111101 in the standard transmission order (LSB first). IEEE 802.5 (Token Ring) and IEEE 802.6, on the other hand, send the bytes over the transmission medium with the MSB first.

There is an important exception in Ethernet data transmission as noted in Figure A.9. Except for the FCS (which has the same byte ordering but a different bit ordering), data on other Ethernet frame fields is transmitted with most significant octet (byte) first, and within each octet, the least significant bit is transmitted first. The byte and bit orders are typical for all fields except the FCS, which is treated as a special 32 bit field rather than as four individual bytes. In the FCS, the high-order bit (MSB) of the sequence is transmitted first.

As discussed above, the 48 bit MAC address (universal or local) is normally represented as a string of 6 octets. The octets are written from left to right, in the order that they are transmitted on the transmission medium, separated by hyphens (−) or colons (:). Each octet of the address is expressed as two hexadecimal symbols. The bits within the octets are transmitted on the transmission medium from left to right.

In the binary representation, the first bit transmitted of each octet on the transmission medium is the LSB of that octet. The I/G address bit is the least significant bit. The leftmost bit of the binary representation (I/G address bit) of a MAC address distinguishes individual from group addresses. The U/L administered address bit is the next bit following the I/G address bit. The U/L bit indicates whether the MAC address has been universally or locally assigned.

A.2.4 Legal Ethernet Frames: Untagged Frames

The minimum and maximum Ethernet frame sizes (in bytes) are determined as follows:

Full minimum frame size = header (14) + CRC (4) + DataMin (46) = 64

Full maximum frame size = header (14) + CRC (4) + DataMax (1500) = 1518

When calculating the frame sizes, the IPG (Inter-Packet Gap), 7 byte Preamble, and 1 byte SFD (Start-of-Frame Delimiter) are not included in the frame size calculations as they are not considered part of the Ethernet frame. The header consists of three parts: 6 byte Destination Address, 6 byte Source Address, and 2 byte Type/Length field

A.2.4.1 Illegal and Ethernet Frames All frames deemed illegal by a receiving Ethernet device are dropped. It is then the responsibility of the higher layer protocols, such as TCP/IP, to notify the sending device that a frame was dropped. The following are illegal Ethernet frames:

- **Runt Frame**

    The minimum Ethernet frame size is 64 bytes, which is equal to 18 bytes Header/CRC plus 46 bytes data. An Ethernet frame that is less than 64 bytes, received by an Ethernet device is deemed illegal and is dropped. This type of illegal frame is referred to as a "runt." The most common causes of runt frames are collisions (in half-duplex networks), buffer underruns, malfunctioning network interface card, or software bugs in a device. In half-duplex networks, such frames are a result of collisions. A receiving Ethernet device is required to discard all runt frames.

    Padding (a smaller Ethernet frame with redundant data) can be used to prevent runts from occurring. If the upper-layer protocol has data to send that is less than 46 bytes, the MAC sublayer adds a sufficient number of zero bytes, 0x00, also known as null padding characters, to the data to meet the minimum Ethernet frame size requirement.

- **Giant Frame**

    The maximum untagged Ethernet frame size is 1518 bytes. An Ethernet frame that is greater than the maximum frame size, received at an Ethernet device is called a "giant." Certain malfunctioning in the physical layer of an Ethernet device may give rise to oversized Ethernet frames. Like runt frames, a receiving Ethernet device is required to discard all giant frames.

- **Misaligned Frame**

    An Ethernet frame must contain an integer number of octets (bytes) (64–1518 for untagged frames). An Ethernet frame that does not contain an integer number of bytes, when received by an Ethernet device is also illegal. An Ethernet device has no way of knowing which bits received are legal. It can only compute the CRC of the frame and also determine if an integer number of bytes are received. A receiving Ethernet device is therefore required to discard such frames.

Some implementations of Gigabit Ethernet (and higher speed Ethernets) support larger frames, known as *jumbo frames*. This type of (nonstandard) frame is not covered in this chapter.

Appendix B

Ipv4 PACKET

B.1 INTRODUCTION

IP (i.e., IPv4 and IPv6) is a network layer protocol that provides connectionless services to upper-layer protocols (e.g., TCP (Transmission Control Protocol), UDP (User Datagram Protocol), Internet Control Message Protocol (ICMP), Internet Group Management Protocol (IGMP), Open Shortest Path First (OSPF)). The connectionless service is implemented with the basic unit of transport being the datagrams (frequently referred to as packets) that contain the source and destination IP addresses of end-user points and other parameters needed for protocol operation.

The IP layer relies on the underlying network infrastructure for transporting the IP datagrams (packets). This means that IP datagrams are encapsulated (in some cases after segmentation into smaller units) by the frames of the underlying network such as Ethernet, ATM (Asynchronous Transfer Mode), and SONET/SD.H (Synchronous Optical Network/Synchronous Digital Hierarchy). Regardless of the physical layer protocol mix, maximum transmission unit (MTU) sizes of interfaces and speed differences of the links in the underlying networks, the IP layer translates these different layer networks into a common logical IP network that is independent of physical characteristics and differences.

The upper-layer protocols such as TCP, UDP, ICMP, IGMP, and SCTP (Stream Control Transmission Protocol) need not be aware of the hardware, encapsulation methods, and other characteristics of the underlying network. Upper-layer

Switch/Router Architectures: Shared-Bus and Shared-Memory Based Systems, First Edition. James Aweya.
© 2018 The Institute of Electrical and Electronics Engineers, Inc. Published 2018 by John Wiley & Sons, Inc.

protocols may expect some levels of quality of service (QoS) from the IP layer during data delivery, such as throughput, delay, and delay variation. These are often referred to as QoS parameters to characterize the nature of the data delivery.

In some cases, the upper layers may pass the QoS expected along with data to the IP layer. The IP layer may support mechanisms to enable it map the QoS parameters to services provided by the underlying network infrastructure. The underlying network may or may not support capabilities to provide the QoS demanded or expected from the upper layer.

The basic IP service is based on "best effort" delivery of data, that is, IP does not guarantee that packets received would be delivered to the next node or final destination, it only tries its best to reach the next node or destination. IP also does not guarantee that packets sent would arrive in the proper sequence or the duplication of packets sent or delivered. Upper-layer transport protocols, such as TCP and SCTP, are responsible for handling sequencing, duplication, and other data integrity issues.

The Internet Protocol version 4 (IPv4), described in IETF (Internet Engineering Task Force) publication RFC 791 (September 1981), is the fourth version or iteration of IP when tracing the development of this protocol (IP). IPv4 was the first version of the protocol to be widely deployed and is the network protocol that currently drives a majority of today's enterprise, service provider, and the Internet.

The previous chapter gives a description of the Ethernet frame format and the different fields that make up a frame. This chapter provides a description of the IPv4 header and its corresponding fields. IPv6, the successor protocol to IPv4, is not discussed in this chapter.

B.2   IPv4 PACKET FORMAT

IP receives data segments from upper-layer protocols (e.g., TCP (protocol number 6), UDP (protocol number 17), SCTP (protocol number 132), ICMP (protocol number 1), IGMP (protocol number 2), OSPF (protocol number 89)) and formats them into IP packets. The IP layer receives the data units from the upper-layer protocols and adds its own header information.

The received data from the upper layer constitute the payload in the IP packet. The IP header contains all the information needed to route the packet through the IP network node to node from the source to the destination (unicast transmission) or destinations (multicast transmission). The IP packets created are then encapsulated by the protocol frames of the underlying network (e.g., Ethernet) as illustrated in Figure B.1.

B.2.1   IPv4 Header

The IPv4 packet header is placed at the front of every IPv4 packet created. The header consists of 14 fields, one of which is optional. The packet is normally

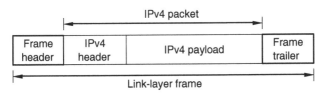

FIGURE B.1 IPv4 Packet carried by a link-layer frame.

20 bytes in length, but sometimes can carry additional options in a variable length field located after the Destination Address field (Figure B.2). The fields in the IP header are formatted with the most significant byte written first (Big Endian), and within the byte, the most significant bit is written first. For example, the Version field in the IP header is located in the four most significant bits of the first byte of the packet. The IP header and the fields it contains are described below:

- **Version:** This 4 bit field specifies the version number of the Internet Protocol used, which also indicates the format of the IP packet header. The header carries other important information other than the version number of the Internet Protocol used. In the context of the protocol discussed here, this field carries a value of 4. The Version field in IPv6 packets contains the value 6.

- **Internet Header Length (IHL):** This 4 bit field specifies the length of entire IP header (including the length of the data in the Options field, when present). The IHL specifies the length of the IP packet header in 32 bit (4 byte) words – that is, the number of 32 bit words in the IP header (Figure B.2). The minimum value for a valid header is 5, which gives a 20 byte ($5 \times 32 = 160$ bits) header (when no options are carried). The maximum IHL value possible is 15, which

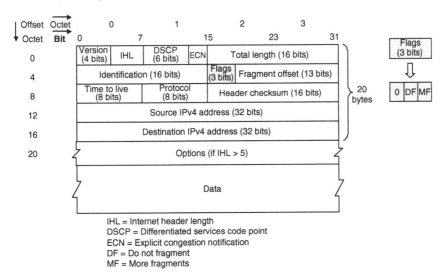

IHL = Internet header length
DSCP = Differentiated services code point
ECN = Explicit congestion notification
DF = Do not fragment
MF = More fragments

FIGURE B.2 IPv4 Packet format.

gives a packet header length of 60 bytes (15×32 bits = 480 bits). The IHL value of 15 allows the header to carry a maximum of 40 bytes of IP options.

- **Differentiated Services Code Point (DSCP):** This 6 bit field is defined in [RFC2474] and occupies part of the obsoleted 8 bit type of service (TOS) field. The six most significant bits (of the obsoleted TOS field) are used for the DSCP and the remaining 2 bits for the ECN (discussed below). The old TOS field (updated in [RFC1349) was used to specify the parameters describing the type of service requested by upper layers. The parameters may be utilized by networks to define how packets from an upper layer client should be handled during transport. The "M" bit was added to the TOS field in [RFC1349].

- **Explicit Congestion Notification (ECN):** This 2 bit field, defined in [RFC3168], carries information about the state of congestion along the route taken by the IP packet from source to destination. ECN is optional and is only effective when the underlying network supports capabilities that allow for the signaling/notification of network congestion state end-to-end. ECN can be used only when the endpoints and the intermediate network support ECN processing capabilities. When used ECN, packets are not dropped (by the intermediate network nodes between the endpoints), but instead the ECN fields in the packets are "marked" with congestion notification information.

- **Total Length:** This 16 bit field specifies the total length (in bytes) of the entire IP packet. This specifies the length of entire IP packet (including the IP header and IP payload). The minimum IP packet length is 20 bytes (i.e., 20 bytes of IP header plus 0 bytes of payload data), and the maximum length is 65,535 bytes (i.e., 2^{16}-1, since only 16 bits are available in this field to specify the total packet length). All network links are expected be able to handle IP packets of at least 576 bytes, but a more typical packet size is 1508 bytes. Note that a fragment of an IP packet is considered a complete IP packet in itself by downstream nodes (since it has its own IP header).

- **Identification:** This 16 bit field is used to identify the fragments of one IP packet from those of another (different) IP packet. If an IP packet is fragmented by a node during transport, all its fragments are given (assigned) the same identification number to enable the destination identify the original/sourced IP packet they belong to. The network node that is the source of the fragments sets the identification field to a value that must be unique for that source–destination pair and for the time the packet will be alive/active in the network.

- **Flags:** This is a 3 bit field that is used during packet fragmentation. Dictated by network and link characteristics, if an IP packet is too large to be transmitted, the 'flags' indicate if the packet can be fragmented or not. A fragment is part of an IP packet but given its own IP header. In the 3 bit flag, the MSB is always set to '0'. The 3 bit Flags field are defined as follows (from high- to low-order bit):
  - **1 bit R (Reserved) Flag:** This should be set to 0.
  - **1 bit DF (Don't Fragment) Flag:** This controls the fragmentation of the IP packet (0 = Fragment if necessary; 1 = Do not fragment).

- **1-bit MF (More Fragments) Flag:** This indicates if a packet (created from a fragment) has additional fragments after it (0 = This is the last fragment; 1 = More fragments follow this fragment).

  When a complete IP packet leaves its source, the source clears the MF (More Fragments) bit to zero and the Fragment Offset field to zero before transmitting the packet.

- **Fragment Offset:** This 13 bit field carries an offset that specifies the exact position of the fragment in the original IP Packet. The offset is measured in units of 8 byte blocks (64 bits) and specifies the offset of a particular fragment starting from the leading end of the original unfragmented IP packet. This information allows the receiving endpoint to properly reassemble the fragments to obtain the original IP packet.

- **Time to Live (TTL):** This 8 bit field is a timer field (that carries a value) used to track the lifetime of the IP packet as it travels through the network. When the TTL value is decremented down to zero at a network node, the packet is discarded. To avoid packets traveling in loops in the network, every packet is sent with the TTL field set to an initial value, which indicates to the network how many routers (hops) the packet is allowed to cross. Typically, the TTL value is set to an integer value indicating the number of hops a packet can cross. At each hop, this integer value is decremented by 1 and when the value reaches 0, the packet is discarded.

- **Protocol:** This 8 bit field specifies the upper-layer protocol that is the recipient of the IP packet payload. It indicates to the IP (Network) layer at the destination host to which next layer protocol the packet payload should be sent, that is, the client protocol. For example, the protocol number of ICMP (Internet Control Message Protocol.) is 1, IGMP (Internet Group Management Protocol) is 2, TCP is 6, UDP is 17, and SCTP (Stream Control Transmission Protocol) is 132.

- **Header Checksum:** This 16 bit field is used to carry checksum value over the entire IP header (only). This allows a receiving IP node to check if the packet is received error-free. The 16 bit checksum is generally performed as a one's complement of the IP header including all IP options. When an IPv4 packet arrives at an IP node (router), the router calculates the checksum of the received header and compares this with the value carried in the checksum field. If the checksum values do not match, the router discards the packet. Modifying any part of the IP header at a node (e.g., TTL) requires recalculation of the IP checksum. Both UDP and TCP have their own checksum fields that allow for errors in the data they receive to be verified for errors. This means errors in the IP packet Data field is to be handled by the upper-layer protocol.

- **Source Address:** This 32 bit field contains the IP address of the sender (or source) of the IP packet. The source IP address may be changed in transit by a network address translation device.

- **Destination Address:** This 32 bit field contains the address of the receiver (or destination) of the IP packet. This address may also be changed in transit by a network address translation device.
- **Options:** This is an optional field (0–40 bytes in length), which is used if the value of IHL is greater than 5 (but not greater than 15) 32 bit words. The Options field may contain values for options such as security, record route, and time stamp. The Options field is variable in size and, when used, increases the overall length of the IP header. This means the IHL will only ever be greater than 5 if there are IP header options. The maximum size of the IP options is obtained as follows: $(15-5) \times 32$ bits $= 320$ bits $= 40$ bytes.
- **Data:** The contents of this field are interpreted based on the client protocol (i.e., value carried in the IP header Protocol field) (Figure B.2). Since the IP header is at least 20 bytes, the maximum payload (data carried in Data field) is limited to 65,535 bytes $- 20$ bytes $= 65,515$ bytes. If a maximum number of IP options are carried, then the maximum payload becomes 65,535 bytes $- 40$ bytes $= 65,495$ bytes. The IPv4 Header is typically followed by an ICMP header, IGMP header, or a Transport Layer (UDP, TCP or SCTP) header, which, in turn is usually followed by client protocol data. The number of bytes in the client protocol data field is the value of the IP Total Length minus the length of any other headers (ICMP or Transport Layer header), minus the value of IHL (usually 20 bytes). A UDP datagram might have a 20 byte IPv4 packet header, an 8 byte UDP Transport Layer header, and 500 bytes of data, resulting in an IP Total Length packet of 528 bytes. A variable length "Padding" is used by the IP layer as a redundant filler to ensure that the data in the IP packet start on a 32 bit boundary. Note that the data portion of the IP packet is not included in the IP header checksum calculation.

Differentiated Services introduces the notion of per hop behaviors (PHBs) that define how traffic (packets) belonging to a particular behavior aggregate (i.e., packets considered to have the same forwarding characteristics) is handled at an individual network node [RFC3140]. PHBs are not indicated directly in IP packet headers; instead, the DSCP values carried in the header can be used in implementing the PHBs in a network node. The 6 bit DSCP field allows for 64 possible DSCP values, but there is no limit on the number of PHBs a node can implement.

Any given network domain can implement its own mechanisms for defining locally the mapping between DSCP values and PHBs. There are standardized PHBs with recommended DSCP mappings, but network operators may choose to implement suitable alternative mappings.

B.3 IPv4 ADDRESSING

A MAC address is an identifier associated with a physical or virtual/logical interface on an Ethernet device or on devices using technologies such as Token Ring and

FDDI. The typical case is the MAC address is permanently stored in a network interface adapter to uniquely identify the interface in a network. MAC addresses belong to and are used at the Data Link Layer, while IP addresses belong to/used at the Network Layer. The IP address of a device interface may change as the device is moved in a network to different IP subnets or VLANs or when it powers up (in a network with a Dynamic Host Configuration Protocol (DHCP) server), but the MAC address remains the same, because it is associated with the device interface.

The 32 bit Source Address field (which contains the IPv4 address of the packet source) may be modified by a NAT (Network Address Translation) device. Also, a source IP address can be all 0s (the unspecified IP address) in certain cases, but it can never be a multicast address (which identifies only a group of multicast receivers). The 32 bit Destination Address field (which contains the IPv4 address of the packet's receiving endpoint) may also be modified by NAT in a reply packet. This returned IP address can be a unicast or multicast IPv4 address, but it can never be all 0s (the unspecified IP address).

For convenience and to make IP addresses more readable, 32 bit IPv4 addresses are often written in the dotted-decimal notation. This consists of splitting the 4 byte address into four groups, each with 1 byte. Each octet of a group is then expressed individually in decimal (taking values from 0 to 255) and separated by periods (.). This allows the 32 bit IPv4 addresses to be conveniently expressed in dotted-decimal notation, in which each octet (or byte) is expressed as a separate decimal number. Within an address octet, the rightmost bit represents 2^0 (or 1), increasing to the left to the first bit in the octet that is 2^7 (or 128). The following are IP addresses expressed both in binary and dotted-decimal formats:

00110011 11001100 00111100 00111011 = 51.204.60.59
11010000 01100010 11000000 10101010 = 208.98.192.170
01110110 00001111 11110000 01010101 = 118.15.240.85

A 32 bit IPv4 address, in general, is organized in two primary parts: the network prefix and the host identifier or number. Based on this, IPv4 addresses are organized as two groups of bits in the address. The first group contains a portion of the most significant bits and constitutes the network address (or network prefix or network block). This part identifies a whole network or subnet.

The remaining part made up of the least significant bits forms the host identifier. This part specifies a particular interface of a host on that network. This distinction between network prefixes (subnetworks or subnets) specified in the IP address is the basis of traffic routing between IP networks and subnets. The network part (prefixes), which can be of variable length, is also the basis on which IP address allocation policies are developed.

All IP (host) interfaces within a single network or subnet have the same network prefix. This picture gets complicated with the use of supernetting (also called prefix aggregation, route aggregation, or route summarization) [RFC1519], which is not discussed in detail in this book. Each individual interface (host) within the network/

subnet also has its own identifier (host identifier/address), that uniquely identifies it in that network, along with its network prefix.

Also, depending on the interface type and the scope of the network, the IP address assigned to it can be either locally or globally unique. Host interfaces that are accessible/visible to other IP nodes or host outside their local network (e.g., email servers, web servers, video servers) must be assigned a globally unique IP addresses. Host interfaces that are accessible/visible only within their local network must be assigned locally unique IP addresses.

The Internet Assigned Numbers Authority (IANA) is the central numbering authority responsible for assigning IP addresses in addition to other related activities such as root zone management in the Domain Name System (DNS), autonomous system number allocation, and so on. IANA is responsible for ensuring that IP addresses given out are globally unique where required. The IANA also has reserved a large IP address space for use by devices interfaces that are not accessible/visible outside their local networks.

IANA allocates (blocks of) addresses to the Regional Internet Registries (RIRs) who then pass on the addresses to their customers (Local Internet Registries (LIRs), Internet service providers, end-user organizations, etc.) to carry out the actual allocation to end users. A local Internet registry receives an address allocation from a regional Internet registry and then assigns parts of the allocation to their customers. Most local Internet registries also operate as Internet service providers.

IPv4 address allocation went through a number of historical changes, from the original ARPANET (Advanced Research Projects Agency Network) address allocation scheme to classful networking, to networking with Variable-Length Subnet Mask (VLSM), and then to Classless Inter-Domain Routing (CIDR).

B.3.1 Original ARPANET Addressing Scheme

In this older IP addressing scheme, the 32 bit IP address was organized into two parts: the network identifier and a host identifier. The network identifier (or network number field) was carried in the most significant (or highest order) octet of the 4-octet IP address. The host identifier (or the local address) was carried in the remaining 3 octets of the IP address (these 3 octets were also called the *rest field*). The network number field (in the most significant 8 bits of an address) specified the particular network a host was attached to, while the local address (rest field) uniquely identifies a host connected to that network. This addressing system allowed the creation of a maximum of 256 unique networks.

This addressing scheme was adequate at that time, because only a few large IP networks existed, one of which was the ARPANET (which was assigned the network number 10). However, with the wide and rapid proliferation of IP networking, the construction of local area networks (LANs), subnets, and large individual IP networks (academic, enterprise, and service provider networks), it became quickly clear that this addressing scheme was inadequate and not scalable for future network growth.

B.3.2 Classful Addressing

In this IP addressing system, the high-order octets of the 4-octet (32 bit) IP addresses are organized in various blocks and defined to create a set of classes of networks. This scheme was to address the limitations of the original ARPANET addressing scheme and to provide flexibility in the number of addresses allocated to networks of different sizes. This IP addressing system defined five address classes: Class A, B, C, D, and E [RFC791]. Each address class, coded in the first 4 bits of the 32 bit IP address, defines either a network size, that is, the number of potential hosts requiring unicast addresses (classes A, B, C) or a multicast network (class D).

The classes A, B, and C are given different number of bits to accommodate the network identifier. The rest of the bits in an address class is used to identify a host within a network using that class. This means each address class has a different maximum number of host identifiers/addresses that can be assigned to potential hosts.

Class D was designated for IP multicast addressing, while class E was reserved for experimental purposes or future applications. During processing of IP packets, a network node would examine the first few bits of the IP address to determine the class of the address and where the actual network identifier starts and ends.

This classful IP addressing scheme is illustrated in Table B.1. This table shows that each IP address class reserves/specifies a different number of bits for its network identifier (prefix) and host identifier:

- Class A addresses designate only the first byte for specifying the network prefix and the remaining 3 bytes for the individual host identifiers.
- Class B addresses designate the first 2 bytes for specifying the network prefix and remaining 2 bytes for the host identifiers.
- Class C addresses designate the first 3 bytes for specifying the network prefix and the remaining 1 (last) byte for host identifiers.

The three IP address classes can be represented in binary format as follows, with an h representing each bit in the host identifier:

Class A: 0NNNNNNN hhhhhhhh hhhhhhhh hhhhhhhh
Class B: 10NNNNNN NNNNNNNN hhhhhhhh hhhhhhhh
Class C: 110NNNNN NNNNNNNN NNNNNNNN hhhhhhhh

Each bit (h) in a host identifier can have a 0 or 1 value. For example, if only 3 bits are reserved for specifying the host identifier, then we have the following possible host identifiers:

000; 001; 010; 011; 100; 101; 110; 111

For each IP address class, if H is the number of host identifier bits, then the maximum number of host identifiers that can be supported by that particular network prefix is 2^H.

Class A Addresses: Maximum number of host identifiers $= 2^{24}$ (or 16,777,216), $H = 24$.

Class B Addresses: Maximum number of host identifiers $= 2^{16}$ (or 65,536), $H = 16$.

Class C Addresses: Maximum number of host identifiers $= 2^8$ (or 256), $H = 8$.

The maximum number of usable addresses for addressing individual hosts in each address class, however, is $2^H - 2$. The minus 2 here is used to account for the predefined all 0s host identifier part used in a network address and the all 1s host identifier part used in a broadcast address. In a classful network, the mask identifying where that network prefix starts and ends is implicitly derived (inferred) from the IP address itself (i.e., from the leading address bits as shown in Table B.1).

In common network practice, the all-zeros (all 0s) in the host identifier is reserved for referring to all hosts in that network (the entire network or subnet). The all-ones (all 1s) in host identifier is used as a broadcast address in the given network or subnet. These reduce the number of identifiers/addresses available for hosts in a network or subnet by 2. The /31 networks (255.255.255.254) are rarely used, typically used only point to point links (RFC3021]. Such a link supports only two hosts (the endpoints), therefore specifying network and broadcast addresses is not necessary

B.3.3 IPv4 Variable-Length Subnet Masks (VLSM)

To address the physical, architectural, size, and management limitations encountered when constructing large networks, network designers often segment large networks into smaller networks/subnetworks. Let us assume three hosts connected to one subnet (Subnet Delta) and three other hosts connected to a second subnet (Subnet Gamma). Combined, these six hosts and the two subnets (Delta and Gamma) form a larger network than the individual ones. If we assume that the entire network is assigned the network prefix (a Class B address) 192.14.0.0, then each of the six hosts will be assigned an IP address that carries this network prefix.

In addition to sharing the same network prefix (i.e., 192.14), the hosts on each subnet share the same subnet (192.14._Delta_.0 and 192.14._Gamma_.0). All hosts in the same subnet must have the same subnet identifier/address. Let us assume that Subnet Delta is assigned the IP address 192.14.125.0, while Subnet Gamma is assigned the IP address 192.14.18.0.

The Gamma subnet address 192.14.18.0 can be expressed in binary notation as follows:

11000000.00001110.00010010.xxxxxxxx

TABLE B.1 Classful Addressing

Class	Leading Bits (Class Identifier)	Size of Network Identifier Field (Bits)	Size of Host Identifier Field (Bits)	Number of Networks	Addresses Per Network	Start Address	End Address
Class A	0	8	24	$128\ (=2^7)$	$16{,}777{,}216\ (=2^{24})$	0.0.0.0	127.255.255.255
Class B	10	16	16	$16{,}384\ (=2^{14})$	$65{,}536\ (=2^{16})$	128.0.0.0	191.255.255.255
Class C	110	24	8	$2{,}097{,}152\ (=2^{21})$	$256\ (=2^8)$	192.0.0.0	223.255.255.255
Class D (multicast)	1110	Not defined	Not defined	Not defined	Not defined	224.0.0.0	239.255.255.255
Class E (reserved)	1111	Not defined	Not defined	Not defined	Not defined	240.0.0.0	255.255.255.255

For this network, the first 24 bits in the 32 bit address are used to identify the subnet, making the last 8 bits not significant for network identification. To identify the Gamma subnet, its address can be expressed as 192.14.18.0/24 (or just 192.14.18/24). The /24 in this notation represents the subnet mask (also written as 255.255.255.0).

In the past, subnets were created based on address classes. A subnet could have 8 (class A; /8), 16 (class B; /16), or 24 (class C; /24) network identifier bits, corresponding to a maximum of 2^{24}, 2^{16}, or 2^8 hosts. As a result, if an entire class B prefix (/16 subnet) is allocated for a network that required only 600 addresses, then 64,936 ($2^{16} - 600 = 64{,}936$) addresses will be wasted.

The introduction of variable-length subnet masks (VLSMs) allowed for IPv4 address spaces to be allocated more efficiently without the wastage seen when using classful addressing [RFC950, RFC1878]. VLSM allows network designers to allocate the number of addresses required for a particular network more precisely. VLSM was developed to allow IPv4 networks to be subdivided conveniently and more efficiently without being constrained by the addressing limitations of classful addressing, particularly address wastage due to unused address space. VLSM provides more flexibility in designing subnets of varying sizes without unnecessary address space wastage. VLSM was the basis on which Classless Inter-Domain Routing (CIDR (discussed below) was developed.

As an example, let us assume a network with the prefix 192.14.18/24 is divided into two smaller subnets, one consisting of 19 hosts and the other of 47 hosts. To accommodate 19 hosts, the first subnet must have 2^5 (32) host identifiers. Assigning 5 bits to the host identifier results in 27 bits of the 32 bit address being left for the subnet identifier. The IP address of the first subnet then becomes 192.14.18.128/27, which can be expressed in binary notation as follows:

11000000.00001110.00010010.100xxxxx

To get the "128" in the above address, the "**100**xxxxx" is converted to "**100**00000" which is equal to the decimal value of 128. The subnet mask /27 covers the first 27 most significant bits of the IP address. For the second subnet of 47 hosts, the network must accommodate 2^6 (64) host identifiers. Assigning 6 bits to the host identifier results in 26 bits of the 32 bit address being left for the subnet identifier. The IP address of the second subnet is therefore 192.14.18.64/26, which in binary notation is as follows:

11000000.00001110.00010010.01xxxxxx

To get the "64" in the above address, the "**010**xxxxx" is converted to "**010**00000," which is equal to the decimal value of 64. Using the larger subnet mask (/24), the network designer is able to assign address bits within it to create the two smaller subnets. With this the allocated address space is used more efficiently.

B.3.4   IPv4 CIDR

As IP networks grew to accommodate more users, it became apparent that many organizations needed larger address blocks than a class C (/24) network provided. These organizations were, therefore, allocated a class B (/16) address block, which in many cases was much larger than their networks required. Also, as enterprise and service provider networks and the Internet grew rapid, the pool of unassigned class B addresses (2^{14}, or about 16,000) was rapidly depleted.

Also, during the early phase of the Internet development, some organizations were allocated address spaces far larger than they actually needed. All these factors among others led to inefficient address allocation and use, as well as poor routing in networks. With the class B addresses seriously on the verge of depletion, a large number of class C address were given out. The large number of the allocated smaller class C addresses we used to create networks (geographically dispersed) resulted in very large routing tables in routers. These smaller networks were designed and dispersed such that they offered little opportunity for route aggregation.

CIDR (which is based on VLSM) was designed to address the limitations of classful addressing [RFC1517, RFC1518, RFC1519, RFC4632]. The classful addressing method of allocating the IP address space (combined with how routing of IP packets is done) constrained networks designed with the smaller address classes from being scalable. The CIDR scheme was developed to allow a network designer to flexibly repartition any address space (without being limited by class boundaries) to create a network with a larger or smaller block of addresses to be allocated to users. Other than replacing the inefficient classful addressing method, CIDR slowed the rapid depletion of IPv4 addresses and the growth in routing table sizes on routers in networks.

In classful addressing as discussed above, the network addresses (prefixes) are written in a field that is one or more bytes in length, resulting in the class A, B, or C addresses shown in Table B.1. IP address allocations were therefore based on the byte boundaries of the 4 bytes of an IP address. A full IP address was considered to be the concatenation of an 8, 16, or 24 bit network prefix and a corresponding 24, 16, or 8 bit host identifier. With this, the smallest address allocation was the class C addresses that carried only a maximum of 256 host identifiers. This was often too small for most organizations. The larger class B addresses carried 65,536 host identifiers, which was often too large to be used efficiently by even large networks/ organizations.

As enterprise and service provider networks grew, it became increasingly apparent that more flexible and efficient IPv4 addressing methods were needed. With CIDR, an address space is allocated to an organization on any address bit boundary, instead of an 1 byte boundaries. CIDR allows a network designer to partition a larger network into various sized subnets, facilitating creating and sizing a network more appropriately to meet the requirements of the organization.

The CIDR notation (which is now the standard way of representing IP addresses) is a syntax for specifying IP addresses and their associated prefixes for the purpose of routing packets in networks. In this notation, a network address or routing prefix in the IP address is written with a suffix indicating the number of bits of the prefix. Examples are, for IPv4, 192.168.18.0/24, and for IPv6, 2001:db8:abcd:0012::0/64.

- The CIDR notation is derived from the network prefix and its size in the IP address (which is equivalent to the number of consecutive leading 1 bit in the network prefix mask).
- The IP address is expressed according to the version of IP used (IPv4 or IPv6) and involves using a separator character, the slash ('/') character, in front of the network prefix size expressed as a decimal number. The CIDR notation is constructed by concatenating the network prefix, a slash character, and number of leading bits of the network prefix expressed as decimal number.
- The IP address constructed may represent the address of a single, distinct interface in a network, or the routing/network prefix of an entire network. The maximum size of the network (i.e., number of distinct host/interface identifiers supported) is derived from the number of identifiers that are possible with the remaining, lower order bits after the network prefix. An IP address followed by a slash (/) and a decimal number (i.e., 127.0.0.3/8) indicates a block of addresses using a subnet mask. The following are important features of the CIDR notation:
  - The address 192.168.100.17/24 represents the IPv4 address 192.168.100.17 and its associated network/routing prefix 192.168.100.0. The prefix is equivalently obtained by applying the subnet mask 255.255.255.0 (which has 24 leading 1 bit) to the address 192.168.100.17.
  - The IPv4 address block 192.168.100.0/22 represents the 1024 IPv4 addresses from 192.168.100.0 to 192.168.103.255 (using a $32 - 22 = 10$ bit host identifier space). This is equivalent to the address in binary notation:

    11000000.00001110.011001**00**.**00000000**

    to

    11000000.00001110.011001**11**.**11111111**

  - The CIDR notation allows for a more compact representation of IPv4 addresses and prefixes than the dot-decimal notation where both the address prefix and the subnet mask are indicated. 192.168.100.0/24 was written in a longer form as 192.168.100.0/255.255.255.0.
  - The number of host/interface identifiers in a subnet (defined by the network prefix (or mask)) can be calculated as $2^{IP\_Address\_Size - Net\_Prefix\_Size}$, in which the "IP_Address_Size" is 32 for IPv4 (and 128 for IPv6). For example, in IPv4, a prefix size of /20 gives $2^{32-20} = 2^{12} = 4096$ host identifiers.

    With the introduction of CIDR, the process of allocating address blocks to organizations is based on the actual short-term and projected long-term needs of the organizations. The CIDR prefix-based method of representing IP addresses and the associated route aggregation properties allow blocks of addresses to be grouped into single routing table entries. This enables routing to be done more efficiently in the Internet. These address groups (commonly called CIDR blocks) each share a common shorter prefix.

CIDR allows for the aggregation of multiple contiguous network prefixes (called route aggregation, route summarization, or prefix aggregation,) for the creation of supernetwork (or supernet) in networks. The resulting supernet has a subnet mask (network prefix) that is smaller than the individual subnet masks (prefixes) used in constructing the supernet. The supernet prefixes are advertised by routers as aggregates, thus reducing the number of entries in routing tables. The advantages of CIDR and route aggregation can be summarized as follows:

- CIDR and supernetting allows the aggregation of routes to multiple smaller networks.
- This results in smaller routing table sizes in routers, as well as saving memory storage space for the routing tables in the routing devices.
- With smaller routing tables, routing decisions (in the routing devices) are simplified (faster look-ups and forwarding).
- With smaller routing tables (and a fewer number of smaller networks visible to the outside networks), routing advertisements to neighboring routers are reduced.
- The use of supernetting allows a network (supernet) to isolate internal topology changes from other outside routers. This can help to improve the stability of the entire network by limiting the propagation of routing protocol traffic outside the supernet after a network link fails internally.
- If a routing device advertises only an aggregate/summarized route (from a supernet) to peer routing devices, then it does not need to advertise any changes in the specific smaller subnets within the supernet (summarized route). The route aggregation can significantly reduce any unnecessary routing protocol updates following a topology change within the supernet. This increases the speed of convergence of network state and allows the overall network to be more stable.

CIDR address blocks are managed by the IANA with assistance from the RIRs. The IANA allocates to the RIRs large, short-prefix CIDR address blocks who are then responsible for distributing these address blocks to their customers. The RIRs (responsible for address management and allocation in each geographic area, North America, Europe, Africa, etc.) divide these short-prefix CIDR address blocks and allocate the subblocks to the LIRs. Similarly, the LIRs subdivide the address blocks they receive and allocate them to their customers. End-user networks receive address prefixes from the LIRs sized according to their network needs.

B.3.5 Reserved IPv4 Addresses

The following are examples of reserved IPv4 addresses:

- **Private Addresses:** These IPv4 addresses can be used within a home, office, campus, company, and enterprise network and are only visible within these networks and not outside [RFC1918]. They are used within such networks when globally routable addresses are not required or obligatory within these networks. Private addresses are not globally managed and delegated by the IANA and RIRs, meaning that they are not allocated to any specific organization and IP packets carrying these addresses cannot be transmitted through the public Internet.

  The following three blocks of the IPv4 address space are reserved by the IANA for private networks:

- **10.0.0.0/8 (255.0.0.0) Addresses:** 10.0.0.0 – 10.255.255.255; 16,777,216 host identifiers (single class A network)

- **172.16.0.0/12 (255.240.0.0) Addresses:** 172.16.0.0 – 172.31.255.255; 1,048,576 host identifiers (16 contiguous class B networks)

- **192.168.0.0/16 (255.255.0.0) Addresses:** 192.168.0.0 – 192.168.255.255; 65,536 host identifiers (256 contiguous class C networks)

  From Table B.1, it can be seen that only parts of the "172" and the "192" address ranges are designated for private use. The remaining addresses in these address classes (Classes B and C) are public and routable on the global Internet.

  Packet with these addresses cannot be routed on the Internet, so such packets are dropped by the routers. In order to communicate with the outside networks, these IP addresses have to be translated to public (routable) IP addresses using a NAT (Network Address Translation) device, or Web Proxy server. A separate range of private addresses was created to control the assignment of the already-limited IPv4 public routable address pool. By using a private address range within a home, office, campus, and similar environments, the demand for routable IPv4 addresses globally decreased significantly. It has also helped delay the exhaustion of routable IPv4 addresses.

- **Loopback Address:** The IANA reserved the IP address range: 127.0.0.0 – 127.255.255.255 (127.0.0.0/8) for use as a host's self-address, loopback address [RFC6890]. The loopback address is also known as localhost address. The loopback (localhost) address is implemented and managed within the host's operating system. An IP packet carrying the loopback address in its source IP address field should never be transmitted outside the host. The loopback address is used within the host to enable the local server and client processes on the host to communicate with each other.

  When a process running on the host generates a packet with destination IP address set to the loopback address, the operating system loops the packet back to the process without any interference from the network interface adapter. Data sent to the loopback address are forwarded by the operating system to a virtual network interface within operating system that turns it around. The loopback address is mostly used for testing (on a single host) how a client–server process works after its implementation.

- **Broadcast Address:** The Destination Address field of an IPv4 packet can carry a special IPv4 broadcast address, that is, 255.255.255.255. Packets carrying this address are never forwarded by routing devices to other networks but remain in the broadcast domains (VLANs or subnets) in which they are sourced.

- **Multicast Addresses:** IPv4 multicast addresses are identified by the four highest order address bits 1110 as shown in Table B.1. This definition originates from the classful addressing scheme where this group of addresses is designated as class D addresses. A multicast address starts with 224.x.x.x and the range is from 224.0.0.0 to 239.255.255.255. The CIDR prefix of the IPv4 multicast address range is 224.0.0.0/4. Address assignments within this range are specified in [RFC5771].

- **Link-Local Addresses:** Link local addresses are defined in [RFC6890] and range from 169.254.0.0 to 169.254.255.255 (address block 169.254.0.0/16). These addresses are not routable and are only used and valid on links such as a point-to-point connection (between two interfaces) or a single local network segment connected to a host interface. This is because a link-local IPv4 address is not guaranteed to be unique beyond the interface on which it is applied. Routers therefore do not forward packets carrying link-local addresses. These addresses cannot be carried in the source or destination IP address fields of packets traversing routers.

  The assignment of link-local addresses may be done manually by a network administrator or automatically through mechanisms and procedures in a host's operating system [RFC3927]. In case a host is not able to obtain an IP address from a DHCP (Dynamic Host Control Protocol) server and it has not been assigned any IP address manually, the host can assign itself an IP address from a range of reserved link-local addresses. In the absence of a DHCP server, the host may randomly choose an IP address from the range of reserved link-local addresses and then check (via ARP) to ensure that no other host has assigned itself the same IP address.

  Once the hosts (on the point-to-point link or connected to the same single network segment) are configured with link-local addresses in the same address range, they can communicate directly (not across a router) with each other. These IP addresses do not allow hosts to communicate with each other when they do not belong/connect to the same physical or logical network link or segment.

302 APPENDIX B: IPv4 PACKET

B.4 ADDRESS RESOLUTION

Address resolution can be done via one of these methods depending on the format of the original address:

- **ARP (Address Resolution Protocol):** In an IP network, ARP is used to map the configured IP address of a device's interface to its corresponding MAC address. ARP is used to obtain the MAC address of an interface whose IP address is already known. Using ARP, a device sends a broadcast packet (ARP request) that is received by all the host interfaces in the broadcast domain (network segment). But only the interface whose IP address is indicated in the ARP request responds to the request by providing its MAC address.

  The MAC address of an interface can be queried given the already known IP address using the ARP for IPv4 or the Neighbor Discovery Protocol (NDP) for IPv6. ARP or NDP is used to translate IP addresses (OSI Layer 3) into Ethernet MAC addresses (OSI Layer 2). A device maintains an ARP table in which it keeps IP address-to-MAC address mappings for the network.

- **DNS (Domain Name System):** DNS is a system through which a device can obtain the IP address of another device whose domain name is already known. Hosts on the Internet are usually known by names, for example, www.myself.com, and not primarily by their IP addresses (which is used for identifying network interfaces and routing). To allow domain names to be used for communications over networks, the former has to be translated (or resolved) to IP addresses and vice versa. The translation between IP addresses and domain names is performed by the DNS server. DNS is a distributed and hierarchical naming system that allows name spaces to be delegated to other DNS servers.

B.5 IPv4 ADDRESS EXHAUSTION

The rapid growth of the Internet and its extensive reach drastically increased the number of devices that needs unique IP to be able to communicate with others. It came to a point that enterprise and service providers could not continue to give their customers globally unique IPv4 addresses, and at the same time could not get new globally unique IPv4 addresses for expanding their networks. In spite of these factors, the service providers were expected to continue to serve both existing customers and accept new customers.

The service provider can accept new customers requiring globally unique addresses if their allocated IPv4 address space can accommodate them. With the 32 bit (4 byte) IPv4 address field, the total IPv4 address space is limited to 4,294,967,296 (2^{32}) addresses. At the start of the Internet, the 32 bit IPv4 address space was considered larger enough for future needs and so there were little concerns about address depletion.

However, the rapid growth of the number of devices (e.g., mobile devices (laptop computers, smart phones, etc.), always-on communication devices (ADSL modems, cable modems), communication-enabled vehicles, and other electronic devices) expanded the demand for extra IPv4 addresses, a situation that was not foreseen at the start of the Internet. As addresses were assigned to all these wide range of new users, the number of unassigned addresses decreased. Furthermore, the rising use of Internet-enabled devices and appliances created great concerns that the public IPv4 address space may eventually be depleted sooner than later. To address these concerns, the following practices were adopted (in addition to CIDR):

- **Private IP Addresses:** Few blocks of IPv4 addresses were designated for private use within private networks (with interfaces not visible to the outside world) so that the demand for public IPv4 addresses can be reduced.

- **NAT (Network Address Translation):** A NAT device is a mechanism (with one or a few public (routable) IPv4 addresses) through which multiple hosts with private IP addresses in a network can communicate with devices in the outside world using public IPv4 addresses. Most devices in a private (residential, campus, or enterprise) network are assigned private IP addresses that are not routable on the Internet. It should be noted that when an IP router receives an IP packet with a private IP address, it does not forward the packet, it drops the packet. So, in order to communicate with devices with public IP addresses, devices in private networks must use an address translation (NAT) service, which translates between public and private addresses. When a device sends an IP packet out of a private network, the NAT replaces the private IP address in the packet with public IP address of the NAT device and vice versa.

- **DHCP (Dynamic Host Configuration Protocol):** DHCP is a protocol through which a host interface in a network is assigned an IP address from a predefined IP address pool the DHCP server maintains. The DHCP server may also provide additional information such as the default gateway (router) for a host, subnet mask, DNS Server IP address, lease time with the assigned IP address, and so on. By using DHCP services, a network administrator can manage the assignment of IP addresses automatically and more efficiently.

- **Proxy Server:** To enable users access the Internet, a network can use a Proxy Server that has a public IP address assigned to interface. All the hosts in the network will send requests to the Proxy Server that will then forward them to a server on the Internet. The Proxy Server acts on behalf of the hosts to send the requests to the server (somewhere on the Internet) and when it receives responses from the server, the Proxy Server forwards them to the client hosts. This has been an effective method for controlling Internet access in private networks and it facilitates the implementation of web-based policies.

- **Unused Public IP Addresses:** These addresses are being reclaimed by the RIRs to be reassigned to new users. By encouraging organizations to implement renumbering of their networks, the RIRs are able to reclaim large blocks of unused IPv4 address space allocated to these organizations during the early development of the Internet. Also, the RIRs are exercising tighter management and control over the allocation of IPv4 address blocks to the LIRs.

The various limitations of IPv4 discovered since its inception spurred the development of IPv6 in the 1990s, which has been in various stages of commercial deployment since 2006. The IETF redesigned the IPv6 addresses with the main goal of addressing the drawbacks of the IPv4 addresses. IPv6 improved many of the functionalities of IPv4, providing changes that improved addressing, security, and configuration and maintenance. An IPv6 packet has a 128 bit address field, large enough to allow very much larger number of devices to be assigned IP addresses.

Currently, a majority of the devices running on the Internet still use IPv4 and it is anticipated that the shift to total IPv6 use is far in the future. This means that IPv4 and IPv6 will coexist for many more years, and this coexistence must be transparent to users of either protocols. A number of mechanisms has been provided by IPv6 [RFC2893, RFC7059], which allow IPv4 and IPv6 to coexist until the Internet shifts to only IPv6:

- Dual IP Stack
- Tunneling (6to4, 4to6, etc.)
- NAT Protocol Translation

It is argued that the best long-term solution to IPv4 address depletion is to move to IPv6. The long-term use of IPv4 is still being debated. IPv6 provides a much larger address space that also allows improved route aggregation across the Internetworks and offers large subnetwork allocations to organizations. Migration to IPv6 is in progress (for communication-enabled vehicles, sensors, Internet-of-Things (IoT) devices, etc.), but complete migration is not expected soon.

Service providers and some enterprises have to make compromises between growing their networks using IPv6 and continuing to serve existing and new IPv4 customers. The technologies and solutions discussed above enable enterprises and service providers to implement mixed IP addressing solutions even as they build IPv6 networks to accommodate new services such as communication-enabled vehicles, sensors, and IoT devices.

B.6   IPv4 OPTIONS

If the IP packet's Internet header length (IHL) is greater than 5 (i.e., it is from 6 to 15) 32 bit words, it means that the packet carries Options and must be processed. The Options field is a variable length field (up to 40 bytes in length) and consists of the following subfields:

- **1 bit C (Copy) Flag:** This indicates if the Options in the IP packet is to be copied into all its fragments (0 = Do not copy; 1 = Copy).

- **2 bit Class Field:** This indicates the class to which the IP Options belong (0 = Control; 1 = Reserved; 2 = Debugging and measurement; 3 = Reserved)
- **5 bit Option Field:** This indicates the type of Options carried in the IP packet. Examples of IP Options are 0 = End of options list; 1 = No operation; 2 = Security; 3 = Loose Source Route (LSRR); 4 = Time stamp; 7 = Record Route; 9 = Strict Source Route (SSRR); 18 = Traceroute; and so on.

The value in the IHL field must include enough extra 32 bit words to hold all the options (plus any padding needed to ensure that the header contains an integer number of 32 bit words). The options field is not often used because of some concerns about network security. Security concerns discourage the use of Loose Source and Record Route and Strict Source and Record Route.

B.7 IPv4 PACKET FRAGMENTATION AND REASSEMBLY

With IPv4, it is possible for routers to fragment packets (split them into multiple smaller units) if required to transmit them on network interfaces that cannot handle larger packets. Packets that are fragmented must be reassembled at the destination node. Fragmentation and reassembly is much complex in IPv4, but the process has been simplified in IPv6.

MTU mismatch occurs when the MTU of the output network interface is smaller than the MTU of the input interface. This mismatch can result in packet fragmentation or discard at the device having the mismatch. It is possible for an IP packet to be fragmented at a routing device, and for the fragments (which are carried in whole IP packets with their own headers) to be again fragmented at another routing device. The Identification (fragment ID) field (16 bits) in the IP header identifies the original packet a fragment belongs to. This information is used by the destination node to reassemble the fragmented packet later. Each original packet is assigned a unique Identification value (fragment ID) and every fragment of that packet is assigned the same Identification value.

In IPv4, any router (including the sender) can fragment an IP packet. Usually, only the destination endpoint reassembles fragments into a complete packet. It is possible for a border firewall (at the edge of a network) to reassemble fragments to obtain the original packets to allow it enforce security filtering rules. In IPv6, packet fragmentation can only be done by the source node, and a fragmentation extension header is carried in the packet header that contains the information needed for the destination node to reassemble complete packets.

The first bit of the IP header Flags field is reserved and all nodes sourcing a packet must set this bit to zero. The second bit (the DF (Don't Fragment) flag), if set (= 1), indicates that any routing device receiving the packet must not fragment it. Instead, if a packet with DF set reaches a routing device whose output interface cannot handle a packet of that size, that packet is dropped (and ICMP Destination Unreachable message is sent to the sender). The DF flag can be used by a source node when it wants to send packets to another node but also wants to determine if the node's interfaces can forward the packet.

The DF flag can be used for Path MTU Discovery [RFC1191], either automatically by features implemented in the host's IP software or manually using diagnostic tools such as ping [RFC792, RFC1122] or Traceroute (IPv4 Option 18) [RFC792]. The third bit (the MF (More Fragments) flag), if MF set indicates that there are more fragments following the particular fragment carrying the MF bit. Packets that have not undergone fragmentation have their MF flag set to zero. Except the last fragment packet, all other fragments of the same packet have their MF flag set.

The 13 bit Fragment Offset field is used by the destination node when reassembling fragmented packets. The offset is measured in 8 byte blocks and is the offset of a particular fragment measured from the front of the original IP packet. The first fragment of an IP packet (which constitutes the start of the packet) has an offset of 0. The last fragment of a packets carries a nonzero Offset field to allow it to be easily differentiated from a packet that has not been fragmented.

The 13 bit Fragment Offset field provides a maximum offset of $(2^{13} - 1) \times 8$ $= 65,528$ bytes, which is greater than the maximum IP packet length of 65,535 bytes (i.e., $2^{16} - 1$) including the 20 byte basic IP header length $(65,528 + 20 = 65,548$ bytes). The maximum IP payload is limited to $65,535 - 20$ bytes $= 65,515$ bytes. Dividing the payload data 65,515 bytes by 8 byte (unit of the offset) results in a maximum of 8189 offset units.

This means Fragment Offset field is limited to maximum of 8189 actual offset units (and not 8191 (i.e., $2^{13} - 1$) when considering the 13 bit Fragment Offset field). An IP fragment carrying an Fragment Offset value set to 8189 (i.e., in the last fragment) could have a maximum payload of 3 bytes:

- Maximum IP packet length 65,535 bytes – minimum IP header length 20 bytes – (8189 offset units × 8 bytes per offset unit) = maximum 3 bytes.

When a router receives a packet, it performs an IP forwarding table lookup using the IP destination address and determines the outgoing interface to use and that interface's MTU. If the packet size is larger than the interface's MTU, and the DF bit is set to 0, the router may fragment the packet. When there is an MTU mismatch, the router may divide the packet into fragments.

The maximum size of each fragment is the MTU of the interface minus the IP header size. The IP header is 20 bytes minimum (without IP options) and 60 bytes maximum (with IP options). The router formats each fragment into an IP packet and with each fragment carrying IP packet modified as follows:

- The total length field of the new IP packet is set to the size of fragment plus the IP header size.
- The MF flag in the packet is set to 1 for all fragment carrying packets except the last one, which is set to 0.

- The fragment offset field in the new IP packet (measured in units of 8 byte units) is set appropriately as described above.
- The IP header checksum field in the new IP packet is recalculated (see discussion below).

For an MTU of L bytes and a basic IP header size of 20 bytes, the fragment offsets would be multiples of $(L-20)/8$. If we take the MTU of the interface to be 1500 bytes and the minimum IP header size of 20 bytes, then the fragment offsets to be carried in the fragment offset field would be multiples of $(1500-20)/8 = 185$, that is, 0, 185, 370, 555, 740, and so on

To identify an arriving IP packet as a fragment, a receiver checks if at least one of the following conditions is true:

- The MF flag is set to 1 (which is true for all fragments of a packet except the last).
- The fragment offset field is nonzero (which is true for all fragments of a packet except the first).

Using the Identification (fragment ID) field, the receiver identifies fragments that belong to the same (original) IP packet so that reassembly can be done. Using both the fragment offset field values and the MF flag, the receiver reassembles the packet from fragments with the same identification field value.

Reassembly may involve placing the fragments in a reassembly buffer, with each new arriving fragment located in the reassembly buffer starting at fragment offset field value × 8 bytes from the front (beginning) of the buffer. When the receiver takes in the last fragment (which has the MF flag set to 0), it can compute the total length of the original data payload by multiplying the offset in the last fragment by 8 and adding the size of the data in the last fragment. After receiving all fragments, the receiver can sequence them in the correct order using their offsets. The reassembled original IP packet is then transferred to the upper-layer protocol for further processing.

If a 2500 byte IP packet is transmitted from a source and is fragmented into chunks (fragments) of 1020 bytes, three fragments can be created as follows:

- **Fragment # 1:** MF Flag = 1; total length = 1020; data size = 1000; offset = 0.
- **Fragment # 2:** MF Flag = 1; total length = 1020; data size = 1000; offset = 125.
- **Fragment # 3:** MF Flag = 0; total length = 520; data size = 500; offset = 250.

In the above fragmentation process, "data size" includes the length of the ICMP, IGMP, or Transport Layer header.

B.8   IP PACKETS ENCAPSULATED INTO ETHERNET FRAMES

The maximum length of IP packets (65,535 bytes) are much larger than the maximum length of Ethernet frames (of 1518 bytes, with a payload of 1500 bytes). This means IP packets larger than 1500 must be segmented and carried in several Ethernet frames. For example, the number of Ethernet frames required to transport an IP packet with maximum size of 65,535 bytes can be calculated as $65{,}535 \div 1500 = 43.69$.

This shows that it takes 44 Ethernet frames to transport one IP packet of maximum size across an Ethernet interface. However, this example does not imply that IP packets are always segmented/fragmented before forwarded over Ethernet. This is because most IP applications are designed to source packets in data blocks smaller than the maximum Ethernet frame size.

B.9   FORWARDING Ipv4 PACKETS

IP forwarding typically involves an IP forwarding table lookup, decrementing the TTL count (by one), recalculating the IP header checksum, encapsulating the IP packet in a Data Link Layer frame, recalculating the Data Link layer checksum, and forwarding the frame to the correct output interface. Forwarding table lookups can be done in hardware, as can the decrementing of the TTL, recalculation of the IP header checksum, and the Data Link layer frame rewrites.

The routing devices also run routing protocols (such as OSPF, RIP, and BGP) allowing them to communicate with other routing devices to generate the information needed to build their routing tables. These routing tables are in turn used to generate the IP forwarding tables that can be used for the IP destination address lookups required to determine the outgoing interface for incoming packets.

B.9.1   CIDR and Routing/Forwarding Table Entries

Allocating blocks of class C addresses was one strategy used to prevent the rapid depletion of class B addresses. However, large class C allocations required many more routing table entries to be maintained in routers. As discussed above, CIDR was introduced to improve both IPv4 address space utilization and routing scalability in the Internet. CIDR allows a block of IP addresses to be aggregated/summarized into a single routing table entry. This consolidation results in a significant reduction in the number of separate routing table entries maintained, particularly, in core routers. The block of IP addresses is consolidated in the routing table entry as follows:

{lowest address in block, supernet mask} or
{lowest address in block, number of common prefix bits}

The start of the address block is the "lowest address in block" and the number of class C addresses in the block is specified by the supernet mask. The supernet mask (or the *CIDR mask*) contains 1s for the common prefix (i.e., part with identical binary values) for all the addresses, and 0s for the parts of the addresses that have different values.

Routes in a CIDR block can be summarized by routing devices in a single router advertisement called an *aggregate*. The networks or subsets that make up a given CIDR block are said to be *more specific* with respect to that CIDR block. The common prefix of the more specific addresses (that make up the CIDR block) is greater than that of the CIDR block itself.

With the introduction of CIDR, routing devices perform IP forwarding table lookups using longest-prefix matching (LPM) searches (also called maximum prefix length match). Currently, all routers support CIDR and use LPM lookups to determine the next hop and outgoing interface for a packet. The CIDR mask is used to determine the number of prefix bits that are to be used in the LPM searches. If there exist multiple routes with different prefix lengths in the forwarding table to the destination, the router selects the route with the longest prefix.

This is because each of the IP forwarding table entries may represent a specific subnet, creating a situation where one IP destination address may match more than one entry. When this happens, the most specific of these matching entries, that is, the one with the longest subnet mask, is referred to as the longest prefix match. It is the forwarding table entry that matches the largest number of leading address bits of the destination address.

An IP address can be checked to see if it is part of a CIDR block. The address is considered to match the supernet/CIDR prefix if the leading (i.e., starting) N bits of the address and the supernet/CIDR prefix are identical. Given the 32 bit IPv4 address and an N-bit CIDR prefix, if 32-N bits do not match, then potentially 2^{32-N} IPv4 addresses could match a given N-bit CIDR prefix. A larger CIDR prefix has potentially fewer address matches, while a smaller prefix has potentially more address matches. This also means that a single address lookup can produce multiple CIDR prefix matches, assuming each prefix has a different length.

Let us consider the following two entries in an IPv4 forwarding table:

192.168.20.15/28
192.168.0.0/16

When a routing device needs to lookup the destination address, 192.168.20.18, both entries in the forwarding table will match this address, meaning both entries in the forwarding table contain this destination address. However, the longest prefix match is found to be the entry 192.168.20.15/28, since the /28 subnet mask is longer than the other mask /16, making the interface corresponding to 192.168.20.15/28 the more specific route.

B.9.2 TTL Update

The 8 bits TTL field in the IP header is used to prevent packets from being forwarded from router to router indefinitely in a network that has routing loops. The TTL was originally intended to be a lifetime limit for a packet in seconds, but it ended up implemented as a maximum lifetime in the number of hops a packet can traverse, that is, a "hop count."

This means that every time a packet traverses a router, the TTL is decremented by 1. If the TTL reaches 0, the packet is dropped. When a packet is dropped, an ICMPv4 nondelivery message (ICMP Time Exceeded message [RFC792) is transmitted to the packet sender. This mechanism governs how the Traceroute command (based on IP Option 18) works.

B.9.3 IPv4 Header Checksum Computation

The IPv4 header checksum, as discussed earlier, is a simple mechanism used in an IPv4 packet to protect (only) the header from data corruption [RFC791]. This 16 bit checksum is calculated only over the IP header bytes and the field in which it is carried is also part of the IP packet header. At each router, an IP packet is modified accordingly and the checksum is recalculated. The packet will be discarded if the calculated checksum does not match the received checksum.

The router must update the IP checksum if it modifies or changes any part of the IP header (such as decrementing the TTL and modifying the DSCP bits). The 16 bit IP header checksum field contains the 16 bit ones' complement of the ones' complement sum of all 16 bit words in the header [RFC1812].

Calculating an IPv4 Header Checksum

- To compute the checksum at a node including the source, the IP checksum field itself is set to zero.
- The IP header is divided into 16 bit words and these words are summed up, and then finally a one's complement of the sum is performed to obtain the IP checksum.
- This means if another node sums the entire IP header, including checksum, the result should be zero if the header is not corrupted.

Verifying an IPv4 Header Checksum

- At any node that receives the packet including the final destination, the IP header is verified to see if it is corrupted. The node *does not* replace (this time) the checksum value in the header with all zeros (i.e., the original/transmitted IP header checksum is not omitted) when verifying the checksum.
- To validate the checksum, all 16 bit words in the header are summed including the transmitted checksum. The result of summing the received IP header,

including the transmitted checksum, should be zero if the header is not corrupted.

- If the result is nonzero, then at least 1 bit in the IP header has been corrupted. However, there are certain multiple bit errors that can cancel out, and hence corrupted IP headers can go undetected.

Since the TTL is decremented by one at each hop (router), the IP header checksum must be recalculated at each hop. References [RFC1141] and [RFC1624] describe how the IP checksum can be computed incrementally (after, for example, a TTL update).

The IPv4 header checksum is eliminated in IPv6. It was argued that the checksums provided in Layer 2 protocols such as Ethernet, ATM (Header Error Control (HEC)), and PPP, combined with the checksums supported in upper-layer protocols such as TCP, UDP, SCTP, ICMP, IGMP, and OSPF were sufficient to make including a separate IPv6 header checksum in IPv6 packets unnecessary.

REFERENCES

[AHMAD89] H. Ahmadi and W. Denzel, A survey of modern high-performance switching techniques, *IEEE Journal on Selected Areas in Communications*, Vol. 7, 1989, pp. 1091–1103.

[AWEYA2000] James Aweya, On the design of IP routers. Part 1: router architectures, *Journal of Systems Architecture*, Vol. 46, 2000, pp. 483–511.

[AWEYA2001] James Aweya, IP router architectures: an overview, *International Journal of Communication Systems*, Vol. 14, No. 5, 2001, pp. 447–475.

[BRYAN93] S. F. Bryant and D. L. A. Brash, The DECNIS 500/600 multiprotocol bridge/router and gateway, *Digital Technical Journal*, Vol. 5, No. 1, 1993, pp. 84–98.

[CISC2TQOS11] Catalyst 6500 Sup2T System QOS Architecture, Cisco Systems, White Paper, April 13, 2011.

[CISC2TQOS17] Cisco Catalyst 6500/6800 Sup2T System QOS Architecture, Cisco Systems, White Paper, February 2017.

[CISC3550DS05] Cisco Catalyst 3550 Series Intelligent Ethernet Switches, Data Sheet, Cisco Systems, 2005.

[CISC3550PRS03] Cisco Catalyst 3550 Series Switches, Cisco presentation, 2003.

[CISC3500XL99] Catalyst 3500 XL Switch Architecture, Cisco Systems, White Paper, 1999.

[CISC6500DS04] Cisco Catalyst 6500 Series Switch, Cisco Systems, Data Sheet, 2004.

[CISCBABY] White Paper, Understanding Baby Giant/Jumbo Frames Support on Catalyst 4000/4500 with Supervisor III/IV, Cisco Systems, Document ID: 29805, March 24, 2005.

Switch/Router Architectures: Shared-Bus and Shared-Memory Based Systems, First Edition. James Aweya.
© 2018 The Institute of Electrical and Electronics Engineers, Inc. Published 2018 by John Wiley & Sons, Inc.

[CISCBQT07] Cisco Systems, Buffers, Queues, and Thresholds on the Catalyst 6500 Ethernet Modules, White Paper, 2007.

[CISCCAT6000] Cisco Systems, Catalyst 6000 and 6500 Series Architecture, White Paper, 2001.

[CISCCAT6500] Cisco Systems, Cisco Catalyst 6500 Architecture, White Paper, 2007.

[CISCCAT8500] Cisco Systems, Catalyst 8500 CSR Architecture, White Paper, 1998.

[CISCQOS05] Cisco Systems, Quality of Service on Cisco Catalyst 6500, White Paper, 2005.

[CISCQoSOS07] Cisco Systems, QoS Output Scheduling on Catalyst 6500/6000 Series Switches Running CatOS System Software, White Paper, Document ID: 10582, April 13, 2007.

[CISCQoSCM07] Cisco Systems, QoS Classification and Marking on Catalyst 6500/6000 Series Switches Running CatOS Software, White Paper, Document ID: 23420, November 16, 2007.

[CISCRST2011] Cisco Systems, Catalyst Switch Architecture and Operation, presentation, Session RST-2011, Cisco Networkers, 2003.

[CISCUBRL06] Cisco Systems, User-Based Rate Limiting in the Cisco Catalyst 6500, Cisco Systems, White Paper, 2006.

[CISCUQoS3550] Cisco Systems, QoS Scheduling and Queueing on Catalyst 3550 Switches, White Paper, Document ID: 24057, May 30, 2006.

[CISCUQSC06] Cisco Systems, Understanding Quality of Service on Catalyst 6000 Family Switches, White Paper, Document ID: 24906, January 30, 2006.

[CISCUQSC09] Cisco Systems, Understanding Quality of Service on Catalyst 6500 Switch, White Paper, February 2009.

[CISCSUPENG32] Cisco Catalyst 6500 Supervisor Engine 32 Architecture, Cisco Systems, White Paper, 2006.

[COBBGR93] Graham R. Cobb and Elliot C. Gerberg, Digital's multiprotocol routing software design, *Digital Technical Journal*, Vol. 5, No. 1,1993, pp. 70–83.

[ETHERALL09] Ethernet Alliance, *Ethernet jumbo frames*, November 12, 2009.

[FROOMRIC03] Richard Froom, Mike Flannagan, and Kevin Turek, Exploring QoS in catalyst, in *Cisco Catalyst QoS: Quality of Service in Campus Networks*, Cisco Press, 2003,

[FORESYSWP96] Fore Systems, PowerHub LAN Switch Architecture, White Paper, December 1996.

[IEEESAEUI48] IEEE Standards Association, Guidelines for 48-Bit Global Identifier (EUI-48).

[IEEESAEUI64] IEEE Standards Association Guidelines for 64-bit Global Identifier (EUI-64) General"

[IEEESAGMAC] IEEE Standards Association, Standard Group MAC Addresses: A Tutorial Guide.

[IEEESAOUICID] IEEE Standards Association, Guidelines for Use Organizationally Unique Identifier (OUI) and Company ID (CID).

[KAROLM87] M. Karol, M. Hluchyj, and S. Morgan, Input versus output queueing on a space-division switch, *IEEE Transactions on Communications*, Vol. COM-35, 1987, pp. 1347–1356.

[MANDYLA04] Lakshmi Mandyam and B. Kinney, Switch Fabric Implementation Using Shared Memory, Freescale Semiconductor, Document ID AN1704, 2004.

[MCKEOW96] N. McKeown, V. Anantharam, and J. Walrand, Achieving 100% throughput in an input-queued switch, *IEEE Transactions on Communications*, Vol. 47, No. 8, 1996, pp. 296–302.

[MENJUS2003] Justin Menga, Layer 3 switching, in *CCNP Practical Studies: Switching (CCNP Self-Study)*, Cisco Press, 2003.

[RAATIKP04] P. Raatikainen, Switch Fabrics, presentation, Switching Technology S38.165, Helsinki University of Technology - Aalto University, 2004.

[RFC791] IETF RFC 791, Internet Protocol, September 1981.

[RFC792] J. Postel, Internet Control Message Protocol, IETF RFC 792, September 1981.

[RFC950] J. Mogul and J. Postel, Internet Standard Subnetting Procedure, IETF RFC 950, August 1985.

[RFC1122] R. Braden, Ed., Requirements for Internet Hosts: Communication Layers, IETF RFC 1122, October 1989.

[RFC1141] T. Mallory and A. Kullberg, Incremental Updating of the Internet Checksum, IETF RFC 1141, January 1990.

[RFC1191] J. Mogul and S. Deering, Path MTU Discovery, IETF RFC 1191, November 1990.

[RFC1349] P. Almquist, Type of Service in the Internet Protocol Suite, IETF RFC 1349, July 1992.

[RFC1517] R. Hinden, Ed., Applicability Statement for the Implementation of Classless Inter-Domain Routing (CIDR), IETF RFC 1517, September 1993.

[RFC1518] Y. Rekhter and T. Li, An Architecture for IP Address Allocation with CIDR, IETF RFC 1518, September 1993.

[RFC1519] V. Fuller, T. Li, J. Yu, and K. Varadhan, Classless Inter-Domain Routing (CIDR): An Address Assignment and Aggregation Strategy, IETF RFC 1519, September 1993.

[RFC1624] A. Rijsinghani, Ed., Computation of the Internet Checksum via Incremental Update, IETF RFC 1624, May 1994.

[RFC1812] F. Baker, Ed., Requirements for IP Version 4 Routers, IETF RFC 1812, June 1995.

[RFC1878] T. Pummill and B. Manning, Variable Length Subnet Table for IPv4, IETF RFC 1878, December 1995.

[RFC1918] Y. Rekhter, B. Moskowitz, D. Karrenberg, G. J. de Groot, and E. Lear, Address Allocation for Private Internets, IETF RFC 1918, February 1996.

[RFC2474] K. Nichols, S. Blake, F. Baker, and D. Black, Definition of the Differentiated Services Field (DS Field) in the IPv4 and IPv6 Headers, IETF RFC 2474, December 1998.

[RFC2475] S. Blake, D. Black, M. Carlson, E. Davies, Z. Wang, and W. Weiss, An Architecture for Differentiated Services, IETF RFC 2475, December 1998.

[RFC2474] IETF RFC 2474, Definition of the Differentiated Services Field (DS Field) in the IPv4 and IPv6 Headers, December 1998.

[RFC2893] R. Gilligan and E. Nordmark, Transition Mechanisms for IPv6 Hosts and Routers, IETF RFC 2893, August 2000.

[RFC3021] A. Retana, R. White, V. Fuller and D. McPherson, Using 31-Bit Prefixes on IPv4 Point-to-Point Links, IETF RFC 3021, December 2000.

[RFC3140] D. Black, S. Brim, B. Carpenter, and F. Le Faucheur, Per Hop Behavior Identification Codes, IETF RFC 3140, June 2001.

[RFC3168] K. Ramakrishnan, S. Floyd, and D. Black, The Addition of Explicit Congestion Notification (ECN) to IP, IETF RFC 3168, September 2001.

[RFC3927] S. Cheshire, B. Aboba, and E. Guttman, Dynamic Configuration of IPv4 Link-Local Addresses, IETF RFC 3927, May 2005.

[RFC4632] V. Fuller and T. Li, Classless Inter-Domain Routing (CIDR): The Internet Address Assignment and Aggregation Plan, IETF RFC 4632, August 2006.

[RFC7042] D. Eastlake 3rd and J. Abley, IANA Considerations and IETF Protocol Usage for IEEE 802 Parameters, IETF RFC 7042, October 2013.

[RFC5771] M. Cotton, L. Vegoda and D. Meyer, IANA Guidelines for IPv4 Multicast Address Assignments, IETF RFC 5771, March 2010.

[RFC6890] M. Cotton, L. Vegoda, R. Bonica, Ed., and B. Haberman, *Special-Purpose IP Address Registries*, IETF RFC 6890, April 2013.

[RFC7059] S. Steffann, I. van Beijnum, and R. van Rein, *A Comparison of IPv6-over-IPv4 Tunnel Mechanisms*, IETF RFC 7059, November 2013.

[SHREEVARG96] M. Shreedhar and G. Varghese, Efficient fair queuing using deficit round robin, *IEEE/ACM Transactions on Networking*, Vol. 4, No. 3, 1996, pp. 375–385.

[TOBAG90] F. Tobagi, Fast packet switch architectures for broadband integrated services digital networks, *Proceedings of the IEEE*, Vol. 78, 1990, pp. 133–178.

INDEX

Switch/Router Architectures: Shared-Bus and Shared-Memory Based Systems, First Edition. James Aweya.
© 2018 The Institute of Electrical and Electronics Engineers, Inc. Published 2018 by John Wiley & Sons, Inc.

IEEE PRESS SERIES ON
DIGITAL AND MOBILE COMMUNICATION

John B. Anderson, *Series Editor*
University of Lund